PLANNING FOR THE PLANET

The Environment in History: International Perspectives

Series Editors: Dolly Jørgensen, *University of Stavanger*; Kieko Matteson, *University of Hawai'i at Mānoa*; Christof Mauch, *LMU Munich*; Helmuth Trischler, *Deutsches Museum, Munich*

ENVIRONMENT AND SOCIETY

For a full volume listing, please see the series page on our website: http://berghahnbooks.com/series/environment-in-history

Planning for the Planet

Environmental Expertise and the International Union for Conservation of Nature and Natural Resources, 1960–1980

Simone Schleper

berghahn
NEW YORK • OXFORD
www.berghahnbooks.com

First published in 2019 by
Berghahn Books
www.berghahnbooks.com

© 2019 Simone Schleper

Library of Congress Cataloging-in-Publication Data

A C.I.P. cataloging record is available from the Library of Congress

Library of Congress Cataloging in Publication Control Number: 2019013213

British Library Cataloguing in Publication Data

A catalogue record for this book is available from the British Library

ORCID:
Simone Schleper: 0000-0002-4906-9813

ISBN 978–1-78920-298-4 hardback
ISBN 978–1-78920-299-1 ebook

Contents

Figures and Tables

Figures

Table

Acknowledgments

At this place I would like to express some words of gratitude to the many colleagues, friends, and family members who have supported me along the way of the writing of this book. I have to thank Maastricht University and the Netherlands Organization for Scientific Research (NWO) (grant number 276-69-002) for allowing me to research and write this manuscript, and the Leibniz Institute of European History, Mainz, for the additional time to complete this publication. I would like to thank Berghahn Books, Dolly Jørgensen, and the two anonymous reviewers for their patience, the excellent guidance, and their valuable suggestions.

Without the help of the archivists, librarians, and former experts who have assisted me in my research, this study would not have been possible. At the different institutions and collections, through email or in interviews, they have dedicated their time and shared their knowledge and recollections. Several of them have welcomed me in their homes, and all of them have made the research time an enjoyable and friendly experience. In particular, I have to thank Gina Douglas, Jennifer Norman, Duncan Poore, Lee Talbot, Hanna and Niels Halbertsma, Jean-Pierre Ribaut, and Daisy Larios.

At the Faculty of Arts and Social Sciences at Maastricht University, I can count many friendly colleagues who have helped me finalize this project. First, I have to thank my Ph.D. supervisors Raf de Bont, investing many hours in the minute reading and the critical dissection of my arguments, and Ernst Homburg, providing oversight, structure, and critical interventions whenever needed. Further, my gratitude extends to several colleagues and mentors from the history department, the Maastricht University – Science, Technology & Society Studies (MUSTS) research cluster, and elsewhere for listening to and reading about my ideas, especially Vincent Lagendijk, Geert Somsen, Pablo Del Hierro, Jens Lachmund, Hans Schouwenburg, Stephen Macekura, Janet Browne, and Everett Mendelsohn, as well as the members of my reading and assessment committees, Christof Mauch, Anna-Katharina Wöbse, Bert Theunissen, and Chunglin Kwa.

Several people I would like to thank for contributing to my wellbeing inside and outside of the office: Marith, Hortense, Karlijn, Valentina, Lisa, Claudia, Daniela, and Anne. Many thanks also to my parents, who provided the necessary advice and encouragement. Most of all, Walter, thank you so much for your support.

At last, I find it important to mention that during the course of this project several of the former experts that feature in my narrative have passed away, some blessed with old age, some after periods of illness: Gerardo Budowski, Duncan Poore, Maurice Strong, and Mostafa Tolba. This book is dedicated to them and the great work that they and their colleagues have done for the conservation of global nature.

Abbreviations

A-OPEC	Arab Organization of the Petroleum Exporting Countries
AWG	Anthropocene Working Group
CCTA	Commission for Technical Co-operation in Africa
CDC	Conservation Development Centre
CGIAR	Consultative Group of International Agricultural Research
CIDA	Canadian International Development Agency
CITES	Convention on International Trade in Endangered Species of Wild Fauna and Flora
ECOSOC	Economic and Social Council
FAO	Food and Agricultural Organization
GATT	General Agreement on Tariff and Trade
GDR	German Democratic Republic
GEMS	Global Environmental Monitoring System
IBP	International Biological Program
IBP/CT	IBP Section for the Conservation of Terrestrial Communities
IBP/PT	IBP Section for the Productivity of Terrestrial Communities
IBRD	International Bank for Reconstruction and Development
ICBP	International Council for Bird Preservation
ICSU	International Council of Scientific Unions
IIAE	International Institute for Environmental Affairs, later IIED
IIASA	International Institute for Applied Systems Analysis
IIED	International Institute for Environment and Development
IMF	International Monetary Fund
IOPN	International Office for the Protection of Nature
IPCC	Intergovernmental Panel on Climate Change
IUBS	International Union of Biological Sciences

IUCN	International Union for Conservation of Nature and Natural Resources
IUPN	International Union for the Protection of Nature, later IUCN
LUC	Land Use Consultants
MAB	Man and the Biosphere Program
MEA	Millennium Ecosystem Assessment
MIT	Massachusetts Institute of Technology
NATO	North Atlantic Treaty Organization
NCS	National Conservation Strategy
NEPA	National Environmental Policy Act, United States
NGO	nongovernmental organization
PEP	Political and Economic Planning
SCIBP	Special Committee of the IBP
SCOPE	Scientific Committee on Problems of the Environment
SDGs	United Nations' Sustainable Development Goals
SSC	IUCN's Species Survival Commission
STS	science and technology studies
UN	United Nations
UNCTAD	United Nations Conference on Trade and Development
UNCED	United Nations Conference on Environment and Development
UNDP	United Nations Development Program
UNEP	United Nations Environmental Program
UNCHE	United Nations Conference on the Human Environment
UNESCO	United Nations Educational, Scientific and Cultural Organization
WHO	World Health Organization
WMO	World Meteorological Organization
WWF	World Wildlife Fund, now Worldwide Fund for Nature

Conserving Global Nature

"Can we keep our planet habitable?" was the question that the *UNESCO Courier* posed to its international readership in January 1969.[1] The issue's cover depicts an oil-covered bird—an auk that had been photographed near a Brittany beach during the Torrey Canyon oil spill of 1967. Its wings covered in sticky black slick, it tries to escape its sad and inevitable fate (figure 0.1). Since the picture was taken, images of oil-covered birds have become symbolic of mankind's exploitive attitude toward nature and people's devastating impact on the planet's wild species. The concern of the Paris-based editors of the *UNESCO Courier*, however, did not involve French waterfowl or wildlife alone. Instead, their message extended to life on Earth in its totality, including their own human conspecifics.

For this reason, the editors of the *UNESCO Courier* had invited scientists and environmentalists from different international organizations to address the state of the global environment. In the January issue, all invited authors echoed similar global concerns and calls for international action. The message was perhaps expressed most clearly by the French zoologist Jean Dorst of the International Union for Conservation of Nature and Natural Resources (IUCN). Stressing the interdependence between the natural and the human world, Dorst explained, "The use of natural resources and nature conservation are two sides of the same problem. . . . The irrevocable disappearance of wildlife . . . would lead to serious disturbances in the overall productivity of the whole biosphere."[2] The solution he proposed was global and based on science. It pertained to the maintenance of the balance between the preservation and use of the resources of the biosphere, the thin envelope around the planet that made life possible. As Dorst expounded, "The pursuit of optimum equilibrium leads logically to the idea of planetary "management" . . . which takes into account the particular uses to which different types of land are designed by their nature."[3] The links between the protection of wild nature and the physical limits to the exploitation of natural resources for the improvement of society's wellbeing were well established at the time. After the heights of postwar economic expansion and reconstruction, the world was confronted with the seemingly lethal combination of pollution, population pressure, and the limits of natural resources.[4]

Figure 0.1. Cover image of the *UNESCO Courier*, January 1969. Reproduced from UNESCO, "Can We Keep Our Planet Habitable?," *UNESCO Courier* 22, no. 1 (1969): 1–44.

The 1960s and 1970s marked a period in which our perception of the natural world and society's place in it was drastically reconceptualized. According to the political historian Jan-Henrik Meyer, our present-day understanding of the environment emerged between the years 1967, the year of the Torrey Canyon spill, and 1973, when a global oil crisis evoked fears of international dependencies and limits to natural resources.[5] Meyer refers to three developments that constituted this change in awareness: the rise of a widespread concern for the environment and its safeguarding, the emergence of new environmental groupings, and the integration of the environment as a political task field. Similarly, the environmental historian Sabine Höhler has called these two decades the beginning of the "environmental age," during which new concerns for global limits to growth emerged alongside new ideas on global environmental management by the means of science and international regulations.[6] Since the early 1960s, public intellectuals in Western society, such as Rachel Carson, Paul R. Ehrlich, Barry Commoner, and Garrett Hardin, spread messages of slow disaster in publications such as *Silent Spring, The Population Bomb, The Closing Circle*, and *The Tragedy of the Commons.*[7] In the same period, the growing public concern about the state of the environment in the global North presented a fertile breeding ground for activist groups and social movements. These grassroots movements exerted additional pressure, demanding the remodeling of existing structures of political organization and mechanisms for decision-making.[8]

With political changes came the quest for environmental expertise. Whereas environmental problems were conceived as global, and solutions seemed to lie in science, the international scientific community was far from working in unison. Adding to the existing institutions concerned with environmental protection, such as IUCN, the number of scientific groups or expert committees concerned with aspects of the environment grew significantly. New approaches from the late 1960s and early 1970s varied widely, ranging from surveys and environmental monitoring, to global networks of research reserves, to local resource development projects in the global South. In 1972, just a few months before the United Nations (UN) Conference on the Human Environment was held in Stockholm, the members of an independent international think tank called the Club of Rome published a much-discussed study: *The Limits to Growth.*[9] This scenario study, based on early computer calculations, predicted the collapse of global society if population growth and patterns of resource extraction continued at the existing rate. Yet the study's implicit suggestion that a single approach could capture and regulate societies and their use of natural resources, irrespective of local socioeconomic conditions and demands, was met with much criticism.[10]

In fact, questions on how to balance the protection of nature with the demands of a growing world population—and whom to trust or burden with the responsibility—remain at the core of environmental policymaking even today. Still, we appeal to politically neutral and universal science to tackle global environmental problems, while solutions and expert roles remain intrinsically linked to political and often locally grounded decisions on how we want to live with nature. This paradox of global environmental expertise and its origins in postwar international science and policymaking is a core theme of this study. This book is concerned with the 1960s and 1970s, two decades in which global environmental ideas were high on the agenda of international science projects, while postcolonial politics and Cold War tensions made truly global solutions well-nigh impossible.[11] At the same time, these two decades marked the beginning of international environmental policymaking and the establishment of many of our present-day institutional mechanisms to investigate and address transnational environmental problems.

Despite the strong entanglement of the domains of knowledge and governance in the environmental policymaking of the period, for a long time historians' accounts have tended to keep the repertoires of science and politics separate.[12] Only recently have the interlinkages between science and politics in global environmental governance found more scholarly attention. Historians of the environment and environmental politics, such as Stephen Macekura, Michael L. Lewis, Bernhard Gissibl, and Etienne Benson, have shown that it is in the entanglement of political agendas and scientific arguments that we need to understand environmental policymaking since 1900.[13] Following a similar approach, this book is concerned with the history of internationally organized and science-based nature conservation since the 1960s.

Nature conservation, concerned with the preservation and management of the living environment in the form of threatened species, endangered habitats, or entire ecosystems, can be considered one of the longest-established, organized ways in which people have been concerned with what would become known as *the environment*. Although notions of what constituted the core activities of conservation have changed over time, as we will see in this study, nature conservation has often been marked off from the strict preservation of nature, aimed at shielding natural landscapes from human impact entirely. Instead, conservation, which emerged in the American context as part of the nineteenth-century National Park movement, has always allowed for a more active management of natural environments through human intervention. Conservation may entail preservation, but can also include a "wise use" of natural resources to some extent.[14] In the European context, in postwar Britain especially, the term "conservation" was closely linked to a more scientific understanding of nature protection and the management of natural resources, land, and landscapes.[15]

A group of leading nature conservationists at IUCN, and their strategies to influence decisions in the emerging field of international environmental politics, present the focal points of this book. Since its foundation in 1948 in Fontainebleau, France, initiated by Julian Huxley of the United Nations Educational, Scientific and Cultural Organization (UNESCO), IUCN has been the biggest and most considerable international and science-based nature conservation organization. Initially called the International Union for the Protection of Nature (IUPN), the organization was created by representatives of twenty-three governments, 126 national institutions, and eight international organizations. In 2019, IUCN's website lists more than 1,300 members, including states, government agencies, scientific and academic bodies, nongovernmental organizations (NGOs), Indigenous Peoples' organizations, as well as business associations.[16] Nevertheless, as the self-proclaimed "best hidden secret in the conservation community," IUCN never reached the renown of its sister organization, the World Wide Fund for Nature (WWF, until 1986 World Wildlife Fund).[17] IUCN's unrenownedness presents perhaps sufficient motivation to dedicate this book to their role in postwar conservation policies. There are, however, other reasons why a focus on IUCN makes for a valuable contribution. This book is concerned with three of them. First, in the postwar period, IUCN continued and consolidated the loosely organized international conservation endeavors dating back to the early 1900s. As an organization, IUCN built on colonial networks of naturalists around the globe; as an epistemic platform, they were in the vanguard of a planetary concern for the global environment. By the 1960s, IUCN could rely on a far-reaching network with scientific members in many parts of the world. Deeply convinced they were the only organization concerned with all aspects of the environment—"all life," "not just those species that attract our eye,"—members offered their advice to policymakers.[18] As such, a focus on IUCN allows investigating a dominant player with a broad understanding of their own competences in the postwar period's international "environmental regime"—the body of institutions, principles, values, and beliefs that governed the conduct of scientific and political actors.[19]

Second, IUCN, insisting on science-based, global conservation guidelines, continued to promote ideas on scientific internationalism stemming from the early 1900s and the interwar period.[20] While UNESCO as a governmental agency with a UN mandate was subject to the changing international politics of decolonization and Cold War tension, IUCN held on to the idea of scientifically neutral and universally valid expert advice throughout the twentieth century. IUCN thus lends itself to an investigation of the postwar history of the "linear model" of expertise in the environmental sciences, which remains a contested issue in the context of present-day climate change policies.[21] This positivist idea that scientists should serve as politically neutral and value-neutral advisers, and that scientific evidence will cause political closure, has been presented as one of the persisting problems for global climate policy by scholars of environmental governance.[22]

Third, IUCN conservationists, as both advocates for nature protection and strong proponents of science-based approaches to environmental problem-solving, occupied an important position between environmental science, politics, and activism. In the miscellaneous international arena of environmental politics, new forms of advocacy, and diverse scientific programs that emerged at the time, IUCN experts were deeply involved in the interplay of scientific thought, new institutional structures for the environment, and the creation of expert roles. In this, IUCN conservationists had to interact not only with different groups of scientists, but also with diplomats, politicians, and heads of states. Such struggles, controversies, and negotiation between different scientific and political groups have played a crucial role in shaping environmental policy, institutions, and expert roles in general.[23] Therefore, a focus on IUCN conservationists allows for the study of the negotiation of environmental policy and a number of related concepts, such as *sustainable development* and *biodiversity*, that aim at reconciling nature protection and the development of natural resources among international organization and in the political sphere.

With a few notable exceptions, IUCN conservationists were scientists: trained naturalists with field experience in zoological and increasingly also botanical projects in remote places, often located in former colonial territories. This book follows IUCN's scientific and executive elites: naturalists affiliated with the organization who held high-ranking positions in IUCN's scientific commissions, executive board, and council during the 1960s and 1970s. At the time, these positions were in virtually all instances occupied by men, trained at leading universities in Europe and North America.[24] The scientists and administrators focused on in this book are the ones who represented the organization at conferences, published in IUCN's name, and were responsible for the scientific line taken by the organization. In particular, this book looks at six of them: the American zoologist Harold Jefferson Coolidge (1904–1985), who served as IUCN's president during the 1960s and early 1970s; Edward Max Nicholson (1904–2003), an English conservationist, head of the British Nature Conservancy, and one of the key figures behind the International Biological Program (IBP); Raymond Dasmann (1919–2002), American biologist and zoologist, who held the position of IUCN Senior Ecologist for most of the 1970s; the German-born Venezuelan agronomist and forestry specialist Gerardo Budowski (1925–2014), who served as IUCN Director General in the late 1960s and early 1970s; Duncan Poore (1925–2016), a British botanist who took over the post of IUCN Senior Ecologist in 1977; and Maurice Frederick Strong (1929–2015), a Canadian self-made businessman, who took on the position of Chairman of IUCN's Bureau in 1977.[25]

The book explores the involvement of these six leading IUCN members and their close colleagues in the shaping of scientific expertise in nature protection during this early period of international environmental concern, when issues such as global environmental degradation, resource development, and the need for nature protection entered discussions in international science and policymaking. In particular, the book seeks to answer three questions. First, it is interested in the changing object and the scientific foundations of nature conservation during the 1960s and 1970s. It therefore looks at how scientists in and outside of IUCN, as well as policymakers, determined what international nature protection was to entail. Building on this first question concerned with the content of nature protection, it then examines how scientific arguments on the scale and scope of conservation were translated into policy strategies for international environmental governance. Therefore, the second question that this book asks relates to how scientists and politicians decided on the methods and the implementation of conservation advice, vis-à-vis other forms of environmental measures. Along with the nature of conservation knowledge and implementation strategies for environmental protection, it looks at discussions on the authority of experts in international environmental politics. The third question that this book investigates is how scientists and policymakers negotiated the sociopolitical role of conservation experts at IUCN in the changing institutional settings of the time.

Exploring the history of global nature conservation through the arguments of IUCN, this book traces the origins of two closely related areas of tension, which continue to dominate environmental policy debates today. The first concerns the conflicting notions of scale between global environmental schemes and calls for locally implemented and community-based projects with a strong focus on the global South. The second polarization is the tension between demands for science-based and politically neutral expertise on the one hand, and the inclusion of a broad range of environmental knowledge and alternative voices on the other hand. By looking at the history of global nature conservation as promoted by IUCN, the book dates the origins of these paradoxes to the very beginning of international environmental politics during the 1960s and 1970s.

Ecosystem Ecology and the Politics of the Environmental Age

With this focus on the interlinkages between the science and politics of global nature conservation, the book closely relates to two topic areas within the field of the environmental humanities. The first concerns the rise of ecology as the main environmental science of the postwar period. The term "ecology" goes back to the nineteenth-century naturalist Ernst Haeckel, originally designating a sub-branch of physiology that entailed both botany and zoology. Ecology has had a variety of interpretations in the life sciences. In the late nineteenth century, ecology as an emerging field of research was supported mainly by antireductionist biologists objecting to the mechanical simplification of nature after the physical sciences, and those who sought to counterbalance the emerging emphasis on laboratory work remote from the field.[26] Since the 1920s, scientists pursuing ecological research studied the interactions of different organisms and their local environments.

Not yet a discipline proper, the research field attained a new boost in the 1960s when ecological thinking became inseparably linked to new ideas on environmental management.[27] Several authors, such as the environmental historians Anna Bramwell and Donald Worster, have associated more holistic ideas in ecology with the radical environmental protests of activist groups in the 1970s.[28] Yet ecology also inspired scientific research projects in which the exploitation of natural resources and the protection of the ecosystems in which these occurred were seen as closely related.[29] Some authors have focused particularly on this kind of system thinking, which linked the use and the safeguarding of natural resources. Historians of science, such as Peder Anker and Joel B. Hagen, have described how the work of 1960s systems ecologists such as the American brothers Howard and Eugene Odum blurred the boundaries between resource management, nature protection, and other forms of land use.[30] This new episteme was inherent to the ecosystem ecology of the 1960s, concerned with understanding the workings of natural systems, and improving and using them.

The term "ecosystem" had been introduced into the field of ecology during the 1930s and 1940s by British and American biologists, such as Arthur Roy Clapham and Arthur Tansley. The term was used as a heuristic tool to describe and study the physical-chemical processes between organisms and their environment. Since the mid-1940s, these closed systems, which could be mapped like food webs, appeared in studies of trophic cycles and the relationships between key stocks of biomass by authors such as Charles Elton and Raymond Lindeman. In the 1960s, in the context of growing concerns about human population growth and resource shortages, these studies gained broader recognition as means to investigate the productivity of different types of ecosystems.[31] Out of these productivity studies emerged in the course of the 1970s the idea—quickly taken up in political debates on environmental protection and economic development—of an existing interlinkage between social and biological systems. The different ideas of interconnectivity within and between ecological and societal systems are particularly important for the narrative of this book, as they built the scientific foundation of the conservation advice provided by IUCN scientists in the period. The notion of closed and manageable systems, determined by the interaction between organisms and their environments, seemed to provide a basic scientific framework and a set of general ecological rules for environmental and social engineering on large geographical scales.[32]

Another important aspect of the ecology of the 1960s and 1970s has been the globalization of the environmental sciences through new planetary thinking and transnational scientific projects. Having emerged from the physical sciences in a postwar military context in North America, "big science" projects were becoming increasingly multidisciplinary and transnational in the 1950s. Big science, although a fuzzy term, is usually used to describe the large-scale scientific inquiries that emerged in the physical sciences out of military-corporate applied research, starting in the 1940s with the Manhattan Project, which involved large and complex machines and forms of organization.[33] The historian of biology David Coleman has discussed how, in the 1950s, a number of influential environmental research groups and laboratories emerged in the United States out of postwar concerns about nuclear effects on the environment, and were funded by the Atomic Energy Commission and the National Scientific Foundation. Research laboratories, such as the Oak Ridge National Laboratory, the Savannah River Ecology Laboratory, or the Brookhaven National Laboratory, combined energy and ecological research, fostered research careers, and helped to establish ecology as a discipline in the American context.[34] The following decade was marked by a growing interest in international cooperation within the new environmental discipline. From the mid-1960s into the 1970s, the first big biology program, the IBP, brought together ecologists and plant and animal biologists from around the world to study both the productivity and vulnerability of natural systems in different regions.[35]

During the 1960s, international ecosystem research furthermore benefited from advances in technology, computer science, and cybernetics, which helped with collecting, sorting, and analyzing global environmental data for the first time.[36] Cybernetics and systems thinking had been developed in the late 1940s by two North American mathematicians, Norbert Wiener and Claude Shannon, who were working on engineering theories of control and communication applicable across the machine-human divide. In this context, information emerged as a key concept that described the messages and feedback loops used by machines as well as organisms to adapt to their environment.[37] Cybernetics and information theory, both seen as universal and interdisciplinarily applicable studies, thus emerged at roughly the same time as the ecosystem idea within the field of ecological research. Not only the systems thinking underlying cybernetics, but also the technological developments that came from advances in communication and information technology, such as computers and satellites, influenced an emerging view of the planet as a closed and manageable environmental system. As the historian of science Fernando Elichirigoity has explained, only these sets of technologies, which allowed monitoring and comparing environmental data on a planetary scale, made possible an apprehension of the environment as a global space.[38]

That the effect of new technologies was not confined to the scientific community alone has been shown by the media historian Robert Poole. The first photographs of Earth seen from space, such as *Earthrise* or *Blue Marble*, taken by the crews of the two Apollo space missions in 1968 and 1972, respectively, transformed public thinking about the global environment.[39] According to authors such as Jeremy Rifkin, Iris Schröder, and Sabine Höhler, in the 1970s all of this resulted in the uptake of a new intellectual paradigm of the global *biosphere*, the biological basis for life on Earth.[40] At the same time, cybernetic thinking in terms of system processes and structures blurred the boundaries between organisms and their environmental life-support systems, and between the natural and the manmade environment. As Höhler has pointed out, powerful metaphors such as "Spaceship Earth" suggested the relevance of not only ecological but also technological expertise to global environmental problems.[41] Discussions during 1960s and 1970s on the scope and focus of conservation advice, as well as environmental expertise more generally, are a core theme of this book.

Alongside the literature concerned with the scientific developments denoting the environmental age, this book adds to the work of authors who have studied the 1960s and 1970s as the birth moment of internationally organized environmental policymaking. This political conceptualization of the biosphere, often drawing on scientific concepts, was reflected in the emergence of international organizations and NGOs concerned with the global environment. The authors of a volume recently edited by Wolfram Kaiser and Jan-Henrik Meyer have shown how during the 1960s and 1970s international organizations, both governmental and nongovernmental, served as important platforms for both political discussions and expert advice.[42] IUCN's own organizational history has been narrated by the British conservation veteran Martin Holdgate, who has mainly focused on internal developments and administrative changes.[43] In addition, several accounts have been concerned with the history of nature protection in intergovernmental organizations with which IUCN conservationists have liaised directly during the period under examination. These include, for instance, Glenda Sluga's international history of the role of UNESCO in environmental problem-solving, or the organizational history of the United Nations Environmental Program (UNEP) by historian of environmental law Bharat Desai.[44] In addition to the latter accounts, which focus on several of the institutions that will play a role in this book, there is a rich historiography on environmental NGOs that were founded in the 1960s and 1970s, and that operated alongside IUCN. The Swiss historian Alexis Schwarzenbach, for example, has reconstructed the early years of IUCN's funding and partner organization, the WWF, established in 1961.[45]

Notwithstanding the universal aspirations of some international organizations, the environment emerged as a topic of global politics at a time when the international community was intrinsically divided. Several authors, such as the historian Thomas Robertson, have particularly looked at how environmental politics during the 1960s and 1970s were complicated by fluctuating relations between the East and the West during the Cold War.[46] Others, such as the political historian Jacob D. Hamblin, have pointed to environmental negotiations across the Iron Curtain.[47] Besides Cold War tensions, another geopolitical conflict, related to the claims to independence and economic assistance by nations in the global South, has been discussed in the context of postwar environmental politics by authors such as Roderick Neumann and John McCormick. Here, the focus has often been on ideas about centralized, global environmental governance and decentralized, local claims to development.[48]

Despite political tensions, the 1960s and 1970s saw the emergence of intellectual compromises and concepts that helped bridge seemingly irreconcilable agendas. In this respect, several important studies have recently emerged across environmental and economic history that have looked at the discipline-spanning concept of *sustainable development*. Concerned with the origins of the notion, Stephen Macekura has traced the convergence of the environmental and economic policy discourse in the second half of the twentieth century.[49] Iris Borrowy, moreover, has pointed out the continued diversity of interpretations that were combined in the concept of sustainable development before it was officially adapted by the World Commission on Environment and Development, also known as the Brundtland Commission, in the mid-1980s.[50] In such policy agreements, scientists often had the role of gatekeepers or negotiators. Alessandro Antonello, for example, has studied the mediating role of individual experts in drafting conservation plans for Antarctic marine living resources during the 1970s.[51]

Scholars of environmental governance have analyzed the wide spectrum of global environmental expertise—ranging across ecology, engineering, and management—which emerged during the two decades this book examines. Authors have moreover highlighted the relevance of geopolitics for environmental problem-solving, for global aspirations often clashed with Realpolitik and existing local traditions in the management of the environment and natural resources. Likewise, they have pointed out the administrative, programmatic, and at times political contexts behind the different organizations and their thematic and geographic foci, demonstrating the differences in their approaches to protecting natural resources and threatened species and the ways they appealed to the public and policymakers.

Despite this rich body of literature on the environmental age, historians have only begun to look at the negotiations on environmental strategies between the promoters of different scientific approaches.[52] This book, then, adds to both literary traditions in the history of environmental sciences and the political history of global environmental governance by shedding light on the thus far little-reflected history of IUCN's role in postwar conservation policy. It focuses on the organization's involvement in negotiations between groups of scientists and policymakers about what global conservation advice and environmental expertise were to entail. First, this book points to the changing scientific foundations of conservation and environmental expertise in general. Throughout the period in question, conservationists involved in IUCN based their claims to universal and globally applicable environmental expertise on ecosystem ecology. Within the ecological reasoning of IUCN elites, the protection and use of natural resources were never separate or irreconcilable objectives, and conservation expertise thus pertained not only to traditional topics such as the protection of threatened species but also to the management of natural resources in general. Based on ecosystem ecology, IUCN conservationists argued for a global conservation approach by linking local sites for protection to the research on universal ecological rules. Yet ecosystem ecology, at the time a young field of enquiry, was not a unified discipline, and various interpretations by conservationists or their opponents existed in parallel. These interpretations differed in their focus on natural or modified systems, functions or parts of systems, and local circumstances or universal rules. Even within single organizations, such as IUCN, different interpretations coexisted, bringing with them substantially different approaches to environmental management, program implementation, and expert roles that needed to be negotiated.

Resulting agreements on ecosystem ecology varied over time. Conservationists in IUCN were constantly adapting their scientific claims. In particular, they remodeled their claims according to the dominant environmental discourse at the UN when this intergovernmental forum became a stage for international environmental politics and diplomacy. With this reorientation, the object of nature conservation also changed. During the 1960s and 1970s, ecosystem ecology as the basis for conservation shifted its emphasis from cybernetics, with an abstract focus on system processes, to a renewed focus on species and wildlife with reference to biological diversity. Yet, discrepancies between local and global aspirations and diverging interpretations of environmental problems and concepts remained. This book aims to bring these controversies and negotiations, often hidden underneath a conflated or shared terminology, to the foreground.

The changing scientific foundations and objects of ecosystem conservation related to the way in which conservationists planned to protect parts of nature and how these plans were to be implemented. Based on a broad perception of ecosystem ecology, IUCN conservationists were actively involved in early debates on the limits of natural resources, resource exploitation, and management, as well as early ideas on sustainable development. Insisting on the universal applicability of ecosystem conservation, they demanded global environmental solutions. Yet, during the 1960s and 1970s, the ideal of universal validity and implementation of conservation guidelines was contested by postcolonial politics and policy changes related to the provision of international scientific and technological assistance. In the 1960s, many former colonial conservation areas, especially in sub-Saharan Africa and Southeast Asia, gained their independence and issued claims for national sovereignty in decisions concerning economic growth and development. Such local claims often seemed difficult to reconcile with calls for global conservation standards. During the course of the 1970s, the integration of nature conservation into the UN system demanded a more decentralized approach. These tensions between globally applicable, universal scientific standards and local priorities and particularities, between binding regulations and flexible interpretations, remained at the core of debates about implementation strategies and methods.

Controversies on the universality of ecological guidelines were linked to decisions on the roles and responsibilities of conservationists in new interorganizational alliances on the environment. With the emergence of environmental politics as a task field, expert roles became an important topic of discussion. Within the new plethora of organizations, conservationists were reformulating their scientific mandate, trying to attain authority as scientific experts while at same time protecting their interest in nature protection within a new geopolitical reality. Throughout the period, IUCN conservationists insisted on their political neutrality as scientific experts. However, decisions on what nature to protect and how to protect it always presupposed a particular and therefore political conceptualization of humankind's place in, and relation to, nature. In this, the global agendas and ideas on scientific neutrality of conservationists, concerned with the protection of nonhuman life, increasingly clashed with those of other groups of scientists or policymakers who promoted a focus on local communities in less developed regions. At the same time, their claim to scientific neutrality was not always compatible with the geopolitical tensions in international postwar politics.

Conceptual and Methodological Approach

Conceptually, the historical research conducted for this book has been informed by the fields of science and technology studies (STS) and the sociology of expertise. Throughout the 1960s and 1970s, questions on what nature protection was to entail, how it was to be implemented, and what the sociopolitical role of the environmental experts was to be were the subjects of disagreement and negotiation. From the very beginning, STS, with its focus on controversies between groups of scientists, and scientists and policymakers, had a strong presence in environmental topics, in which it highlighted the variety of scientific and public opinions.[53] Scholars of STS have been concerned with debates about the soundness of scientific claims and knowledge, and have demonstrated how controversies about science and expertise can function as a tool to better understand the different standpoints behind a scientific compromise or policy decision.[54] According to American sociologist of science Sheila Jasanoff, controversies therefore allow the study of policy agreements not just in terms of success or failure. Many controversies reveal the process of negotiations between opposing interpretations, claims, and aims. Controversies show how scientific identities and solidarities between scientists emerge, as well as the development of both cognitive and normative aspects of scientific theory and practice.[55] It is for this reason that the sociologist Ronald Giere has called controversies the "natural laboratory for studying operations of science and technology and their interactions with the surrounding society."[56] Three controversies between IUCN elites, fellow scientists, and policymakers underlie the narrative of this book.

The three controversies chosen—IUCN members' participation in the IBP (1964–1974), IUCN's representation at the United Nations Conference on the Human Environment (1970), and the drafting of IUCN's *World Conservation Strategy* (1980)—describe a chronological sequence and can thus help us understand how changes in the organizational structures and the argumentative strategies of science-based, internationally organized conservation efforts manifested themselves over time. Equally important are the different natures of these moments. In each controversy, IUCN elites negotiated their science and expert role vis-à-vis other types of environmental knowledge and experts.[57] In the two decades in question, however, with the rise of environmental politics and environmental coordination within the UN, the function and responsibilities of conservation experts changed drastically. Each of the controversies therefore has a different emphasis, allowing the study of the relationship between IUCN elites and the scientific community concerned with the global environment in the postwar period, IUCN's position in the realms of the development politics and diplomacy of the early 1970s, and IUCN's role in the negotiations on scopes of competences between international organizations of the 1980s.

These different types of negotiations often occurred behind the closed doors of expert meetings or in private exchanges between scientists or politicians. Therefore, I draw on the work of sociologist of science Stephen Hilgartner, who has made explicit the difference between public and closed debates. According to Hilgartner, in science policy, the deliberation of arguments within and between groups of experts often happens in meetings and correspondence that remain hidden from the public—"backstage," as he calls it.[58] Respectively, I study the discussions between conservationists and policymakers, looking at both scientific lines of argumentation in front-stage publications for peers, politicians, and the public, as well as exchanges between experts behind the scenes in the form of correspondence, memos, drafts of policy documents, and reports of closed meetings, in order to reflect the issue from all sides.

Yet, while Hilgartner has focused on scientific controversies as such, I am interested in the interlinkage of scientific and policy debates.[59] Overall then, this book is inspired by the notion that policy problems are "hybrids of the scientific and the political."[60] In particular, I draw on the idea of *co-production* as coined by Jasanoff.[61] The concept describes the mutual shaping of science and governance—in other words, the close relationship between our ways of understanding the world and the way we want to live in it. Acknowledging that science is socially constructed, co-production grants a large role to the institutional and social environments in which expertise and approaches to environmental governance, or implementation strategies, evolve, making it particularly relevant for my study on environmental expertise in international organizations. As Jasanoff explains, besides controversies regarding the validity of scientific claims, scientific advisors in politics are involved to a large degree in negotiating different policy options that in themselves carry political weight.[62] Similarly, environmental policy scholar Roger Pielke has shown that when experts are stakeholders in the policy process—as conservationists were in environmental policymaking—they usually engage in policy options and implementation strategies.[63] In addition to negotiations on scientific content, I therefore look at the implementation strategies suggested by conservationists and their opponents. Another important dimension of the ways in which experts engage in the policy process is what historians of science such as Evert Peeters have called "expert performance."[64] This includes the self-fashioning of experts as scientific authorities. The process in which these roles are constructed or accepted plays a crucial role in decisions on expertise, as Jasanoff and Hilgartner have pointed out.[65] After the science of conservation and suggested implementation strategies for the protection of nature, expert roles then form my third research focus. As I will show, discussions on conservation advice, its use, and the expert roles conservationists at IUCN could fulfill were inherently interlinked in the lines of argument of both conservationists and their negotiation partners.

To detect, disentangle, and explain the different conservation controversies, this book draws on a wide array of sources. The first body of studied material includes published documents on organizational projects and agendas; reports of meetings, workshops, conferences; and programmatic pamphlets in which scientific groups argued for the need to protect nature and presented scientifically sound ways of doing so. Aside from conference proceedings and policy documents, many scientists published their scientific opinions in individual publications that linked scientific conservation to the larger questions of humankind's relation to the environment, resource usage, environmental justice, and global politics, and these, too, have been useful in this study. Additionally, in order to understand the developments pertaining to the self-understanding of science-based conservation as a field, I have systematically studied several scientific and organizational periodicals. These include the *IUCN Bulletin* (issues from 1970 until 1985) and *Environmental Conservation* (issues from 1975 until 1985), which are available in various libraries. Further, I examined periodicals published by the organizations that IUCN conservationists tried to partner up with, such as the *UNESCO Courier* (issues from 1960 until 1985) and UN Food and Agricultural Organization's (FAO) major annual public report, *The State of Food and Agriculture* (issues from 1960 until 1985). This body of published sources has helped me to understand the ways in which conservationists and other scientific groups have addressed different audiences, defining their role for their scientific peers and for policymakers, as well as for the concerned public.

Next to these publicly accessible and official documents, the largest portion of the empirical material collected came from unpublished reports of conservation projects or surveys, minutes of meetings, drafts of publications, interview transcripts, and correspondence between different scientists and policymakers. These unofficial deliberations on the science of nature protection, implementation strategies, and claims to expert authority revealed where the controversies lay, where compromises could be made, and where differences persisted. I worked extensively with several archival collections, all of which hold a combination of the private and professional papers of former key conservation figures and high-ranking administrators. I spent the most time working with the papers of Harold Jefferson Coolidge at the Harvard University Archives, as well as the papers of Maurice Strong and Peter Thacher, both held by the Harvard Center for the Environment and Sustainability. Significant material was collected from the papers of Max Nicholson held at the Linnean Society and the Royal Geographical Society in London, at Aberdeen University, and at the Alexander Library for Ornithology in Oxford. Other important collections include the IBP papers at the Royal Society based in London, the IUCN Library in Gland, the papers of British IUCN conservationist Richard Fitter at the Weston Library, Oxford, the papers of British IUCN and WWF conservationist Sir Peter Scott at Cambridge University, and the small collection of UNEP and IUCN papers held by the library of the Royal Tropical Institute in Amsterdam. Where possible, these primary sources were complemented with semistructured interviews or written communication with relevant historical actors.[66]

The Chapters

Based on the conceptual framework and source work described above, the rest of this book proceeds as follows. Chapter 1, "Old Hands, Pastures New:

International Nature Conservation and the Environmental Age," examines the challenges that the well-established network of IUCN conservationists faced in the 1960s and 1970s as they entered a new intellectual and political discourse on the global environment. It draws attention to the negotiation between newly emerging environmental approaches, policy discourses, and groups of experts. It shows that IUCN conservationists had a long tradition of responding to new kinds of ecological research. However, among the multitude of new environmental approaches, they faced new challenges of non-ecological alternatives. The chapter also shows how IUCN conservationists, insisting on their scientific neutrality, were able to circumvent upcoming Cold War tensions that intergovernmental organizations faced. Nevertheless, their long-established network was challenged by criticism on the authority of Western expertise, linked to decolonization and development, demanding the inclusion of more experts from the global South.

The second chapter, "Classifying Ecosystems: The International Biological Program, 1964–1974," looks at the reorganization of the field of conservation itself. This chapter discusses the controversy between different groups of high-level conservationists affiliated with IUCN about what the IBP meant for the future of conservation. One group of IUCN experts that formed around the British conservationist Nicholson pushed for the scientific and political recognition of their expertise by linking conservation to the emerging field of systems ecology and by using the IBP as a vehicle for top-down implementation. At about the same time, a second group that centered on IUCN Senior Ecologist Dasmann took shape at the executive center of IUCN. In contrast to Nicholson and his colleagues in the IBP, this second group based its notion of environmental expert roles on more descriptive ecological studies—in particular, landscapes as practiced by UNESCO. Their respective visions entailed different ecological philosophies as well as different political ideologies regarding the global implementation of conservation. I demonstrate that at the end of the IBP, scientific conservation within IUCN was based on a compound blend of the science and implementation strategies as brought forth by the two groups. In the years to come, IUCN conservationists would continue to promote ecosystem ecology and ecosystem conservation as the scientific endeavors that linked their work more closely to the regional ecological work of UNESCO and other UN agencies.

The third chapter, "Expertise and Diplomacy: Systems Politics at the UN Stockholm Conference, 1972," explores what the advent of UN environmental politics meant for the role of IUCN conservation experts. Looking at the nature diplomacy of IUCN conservationists, the chapter draws attention to the diverse interpretations of the global environment and the different disciplinary approaches to solving environmental problems around 1970. As a reaction to emerging pessimistic voices related to Earth's limited resources and physical boundaries to growth, the conference organization began to formulate environmental problems in social and economic terms. For UN conference organizers, the best approach for environmental problem-solving involved the management and technological improvement of resource trade and distribution, not ecology. Within the UN's approach, the conservation expertise of scientists such as IUCN President Coolidge or Director General Budoswki was narrowly defined as wildlife preservation and the creation and maintenance of national parks. This demarcation granted very limited space to conservation and land-use measures based on ecological knowledge, entailed in the broader endeavors IUCN conservationists had in mind.

The fourth chapter, "Nature's Value: The Fault Lines in the World Conservation Strategy, 1975–1980," examines a controversy between conservationists at IUCN and experts in UNEP, the UN's new environmental agency. In the aftermath of the Stockholm Conference, ideas on environmental problems as pertaining to both nature and society became increasingly shared between different types of organizations inside and outside the UN system. This convergence of environmental discourses around the concept of sustainable development, however, only camouflaged persisting controversies between Mostafa Tolba and Peter Thacher at UNEP and acting Director General Poore and others at IUCN when it came to conservation approaches on the ground, their institutional foundation, and the roles of conservationists in solving environmental problems. In this respect, the *World Conservation Strategy* did not present a new, unifying conservation paradigm but rather, this chapter proposes, the final document manifested two different organizational profiles, including different scientific approaches, expert networks, and value-making practices, resulting in a continued conflict between those defending the innate value of biological diversity and those stressing the economic value of ecosystem processes.

The concluding chapter, "Global Nature Conservation and Environmental Expertise, 1960s–Present," links the main findings from the four chapters back to discussions on the legacy of the environmentalism of the 1960s and 1970s. It highlights the emergence of two polarizations that may help us understand the limited success of IUCN conservationists' appeal for global, ecosystem ecology–based schemes. These areas of tensions continue to exist on the one hand between a planetary concern and local development, and on the other hand between scientific neutrality and authority, and inclusive and politically sensitized environmental decision-making. By looking at conservation and environmental policy efforts of the 1980s and 1990s, including the development of national conservation strategies during the 1980s, the second *World Conservation Strategy* called *Caring for the Earth* from 1991, the UN Conference on Environment and Development, held in Rio in 1992, as well as the Convention on Biological Diversity issued at the conference, I show how these tensions remain relevant beyond the early environmental years, continuing to shape international environmental discussion even today.

Notes

1. At the time, the *UNESCO Courier* was printed in Arabic, English, French, German, Hebrew, Hindi, Italian, Japanese, Russian, Spanish, and Tamil, and read in more than 125 countries.
2. Jean Dorst, "A Biologist Looks at the Animal World (Beast and Men)," *UNESCO Courier* 22, no. 1 (1969): 17.
3. Ibid., 17–18.
4. Mohammad Taghi Farvar and John P. Milton, *The Unforeseen International Ecologic Boomerang: Conference on the Ecological Aspects of International Development* (New York: American Museum of Natural History, 1969); Patrick Kupper, "Die '1970er Diagnose.' Grundsätzliche Überlegungen zu einem Wendepunkt der Umweltgeschichte," *Archiv für Sozialgeschichte* 43 (2003); Kai F. Hünemörder, *Die Frühgeschichte der globalen Umweltkrise und die Formierung der deutschen Umweltpolitik (1950–1973)* (Stuttgart: Franz Steiner Verlag, 2004); also see e.g., Rex Weyler, *Greenpeace: How a Group of Ecologists, Journalists, and Visionaries Changed the World* (New York: Rodale Books, 2004); Thomas Robertson, *The Malthusian Moment: Global Population Growth and the Birth of American Environmentalism* (New Brunswick: Rutgers University Press, 2012).
5. Jan-Henrik Meyer, "Sammelrezension [Reviews]: Where Did Environmentalism Come From? Rome, Adam: *The Genius of Earth Day. How a 1970 Teach-In Unexpectedly Made the First Green Generation.* New York 2013 / Hamblin, Jacob Darwin: *Arming Mother Nature. The Birth of Catastrophic Environmentalism.* New York 2013 / Zelko, Frank: *Make it a Green Peace!. The Rise of a Countercultural Environmentalism.* New York 2013," *H-Soz-Kult,* 21 July 2016, http://www.hsozkult.de/publicationreview/id/rezbuecher-22483.
6. Sabine Höhler, *Spaceship Earth in the Environmental Age, 1960–1990* (London: Pickering & Chatto, 2015), 3; also see Joel B. Hagen, "Teaching Ecology during the Environmental Age, 1965–1980," *Environmental History* 13, no. 4 (2008); Frank Uekötter, *Umweltgeschichte im 19. und 20. Jahrhundert* (Munich: Oldenbourg Verlag, 2007), 73ff. For an in-depth study of the concept in postwar Western society, see Paul Warde, Libby Robin, and Sverker Sörlin, *The Environment: A History of the Idea* (Baltimore: John Hopkins University Press, 2018).
7. Paul R. Ehrlich, *The Population Bomb* (New York: Buccaneer Books, 1968); Barry Commoner, *The Closing Circle: Man, Nature, and Technology* (New York: Knopf, 1971); Garrett Hardin, *The Tragedy of the Commons* (Washington, DC: American Association for the Advancement of Science, 1968); Rachel Carson, *Silent Spring* (Boston: Houghton Mifflin Harcourt, 1962).
8. E.g., Hünemörder, *Die Frühgeschichte der globalen Umweltkrise.*
9. Donella H. Meadows et al., *The Limits to Growth: A Report for the Club of Rome's Project on the Predicament of Mankind* (New York: Universe Books, 1972).
10. Other lines of criticism include the lack of data on which the model was built, as well as the presumed neutrality and lack of bias in computer models. For a detailed discussion

on how *The Limits to Growths* reflected a common sentiment of neo-Malthusian gloom and doom, see John McCormick, *Reclaiming Paradise: The Global Environmental Movement* (Bloomington: Indiana University Press, 1991), 75ff.

11. Timothy Luke has shown how discussions on environmental governance, including methods and strategies for its implementation, were an important realm of political power struggles during the 1960s and 1970s. Timothy W. Luke, "On Environmentality: Geo-Power and Eco-Knowledge in the Discourses of Contemporary Environmentalism," *Cultural Critique* 31 (1995): 149–51.

12. An observation of this kind has been made for the science and politics of nature conservation of the first half of the twentieth century: e.g., see Raf De Bont and Geert Vanpaemel, "Editorial Introduction to Special Section, 'The Scientist as Activist: Biology and the Nature Protection Movement, 1900–1950,'" *Environment and History* 18, no. 2 (2012). For recent calls to study politics and science as intertwined, see the work of STS scholars and STS-inspired environmental historians such as Bruno Latour, *Politics of Nature: How to Bring the Sciences into Democracy* (Cambridge, MA: Harvard University Press, 2004); Dolly Jørgensen, Finn Arne Jørgensen, and Sara B. Pritchard, *New Natures: Joining Environmental History with Science and Technology Studies* (Pittsburg: University of Pittsburgh Press, 2013); Sheila Jasanoff, ed., *States of Knowledge: The Co-Production of Science and Social Order* (New York: Routledge, 2004).

13. These authors have looked at vastly different subject matters, ranging from environmental science and politics in the history of postwar economics, across North American national ideologies of wilderness, and the role that science and politics played in the German East African colonies, to the role played by land use politics in scientific nature conservation measures after Ugandan independence. Still, all of them discuss to some extend the use of scientific arguments in natural resource and environmental politics between different groups of stakeholders. Stephen Macekura, *Of Limits and Growth: The Rise of Global Sustainable Development in the Twentieth Century* (Cambridge: Cambridge University Press, 2015); Michael L. Lewis, ed., *American Wilderness: A New History* (Oxford: Oxford University Press, 2007); Bernhard Gissibl, *The Nature of German Imperialism. Conservation and the Politics of Wildlife in Colonial East Africa* (New York: Berghahn Books, 2016); Etienne Benson, "Territorial Claims: Experts, Antelopes, and the Biology of Land Use in Uganda, 1955–75," *Comparative Studies of South Asia, Africa and the Middle East* 35, no. 1 (2015).

14. E.g., see Richard West Sellars, *Preserving Nature in the National Parks: A History* (New Haven: Yale University Press, 2009), 43ff.

15. David Evans, *A History of Nature Conservation in Britain* (London Routledge, 1992), 10.

16. IUCN, "Members," n.d., accessed 15 March 2019, https://www.iucn.org/about/members/iucn-members.

17. These were the words of a senior IUCN member during the concluding discussion at the conference "Experts and the Global Environment in the 20th Century: A History of Coproduction and Negotiation," held at Maastricht University in January 2016.

18. Dasmann in *Second World Conference on National Parks: Yellowstone and Grand Teton National Parks* (Morges: IUCN, 1972); Martin W. Holdgate, *The Green Web: A Union for World Conservation* (Gland: IUCN, 1999), 40.

19. For a detailed discussion of the use of the regime concept in international environmental policymaking, see Oran R. Young, "Improving the Performance of the Climate Regime: Lessons from Regime Analysis," in *Oxford Handbook on Climate Change and Society*, ed. John S. Dryzek and Richard B. Norgaard (Oxford: Oxford University Press, 2011), 625–638.

20. For the internationalization of the scientific community in the life sciences and its role in peace diplomacy in the first half of the twentieth century, see Elisabeth Crawford, *Nationalism and Internationalism in Science, 1880–1939: Four Studies of the Nobel Population* (Cambridge University Press, 2002); Nikolai Krementsov, *International Science between the World Wars: The Case of Genetics* (New York: Routledge, 2004).

21. Roger A. Pielke, *The Honest Broker: Making Sense of Science in Policy and Politics* (Cambridge: Cambridge University Press, 2007); Silke Beck, "Moving Beyond the Linear Model of Expertise? IPCC and the Test of Adaption," *Regional Environmental Change* 11, no. 2 (2011).

22. Sheila Jasanoff, "A New Climate for Society," *Theory, Culture & Society* 27, no. 2–3 (2010); Kari De Pryck and Krystel Wanneau, "(Anti)-Boundary Work in Global Environmental Change Research and Assessment," *Environmental Science and Policy* 77, no. 1 (2017).

23. Looking at the history of climate policymaking at the UN level, Clark Miller has shown how scientific bodies have often invested much work into a specialized language and organizational structure to underline their position as distinct from politics, while it was the very participation in political discourse and discussion that elevated them into expert roles. Clark A. Miller, "Climate Science and the Making of a Global Political Order," in *States of Knowledge: The Co-Production of Science and Social Order*, ed. Sheila Jasanoff (New York: Routledge, 2004), 48ff., 60ff.

24. I regret that this has turned the work into an almost all male narrative and hope that future research will continue to investigate the scientific work done by female companions and conservationists in their own right, following examples such as Anna-Katharina Wöbse, "Lina Hähnle (1851–1941): Vogelschutz in drei Systemen," in *Spurensuche: Lina Hähnle und die demokratischen Wurzeln des Naturschutzes*, ed. H.-W. Frohn und Jürgen Rosebrock (Essen: Klartext Verlag, 2017), 35–56; "Phyllis Barclay-Smith: Eine eigensinnige Naturschützerin," in *Vordenker und Vorreiter der Ökobewegung*, ed. Udo E. Simonis (Stuttgart: Hirzel, 2014), 103–110.

25. Short biographies of these individuals can be found in the section "Expert Biographies" in the appendix.

26. For an extensive discussion of the origins and the formation of ecology as a scientific field and discipline, see Robert McIntosh, *The Background of Ecology: Concept and Theory* (Cambridge: Cambridge University Press, 1986).

27. E.g., Joachim Radkau, *Die Ära der Ökologie: Eine Weltgeschichte* (Munich: C. H. Beck, 2011); Wolfram Kaiser and Jan-Henrik Meyer, "Introduction: International Organizations and Environmental Protection in the Global Twentieth Century," in *International Organizations & Environmental Protection: Conservation and Globalization in the Twentieth Century*, ed. Wolfram Kaiser and Jan-Henrik Meyer (New York: Berghahn Books, 2017); Ludwig Trepl, *Die Idee der Landschaft: Eine Kulturgeschichte von der Aufklärung bis zur Ökologiebewegung* (Bielefeld: Transcript Verlag, 2014), 231ff.

28. Anna Bramwell, *Ecology in the 20th Century: A History* (New Haven: Yale University Press, 1989); Donald Worster, *Nature's Economy: A History of Ecological Ideas*, Studies in Environment and History (Cambridge: Cambridge University Press, 1985).

29. Peter J. Bowler and Iwan Rhys Morus, *Making Modern Science: A Historical Survey* (Chicago: University of Chicago Press, 2005), 223.

30. Elena Aronova, Karen Baker, and Naomi Oreskes, "Big Science and Big Data in Biology: From the International Geophysical Year through the International Biological Program to the Long Term Ecological Research (LTER) Network, 1957–Present," *Historical Studies in the Natural Sciences* 40, no. 2 (2010): 200; Robert McIntosh, *The Background of Ecology: Concept and Theory* (Cambridge: Cambridge University Press, 1986); e.g., Andrew C. Isenberg, *The Oxford Handbook of Environmental History* (Oxford: Oxford University Press, 2014); Joel B. Hagen, *An Entangled Bank: The Origins of Ecosystem Ecology* (New Brunswick: Rutgers University Press, 1992), 144ff.

31. David G. Raffaelli and Christopher L. Frid, eds., *Ecosystem Ecology: A New Synthesis* (Cambridge: Cambridge University Press, 2010), 4ff.

32. Andrew Jamison, for instance, tracing the history of environmental systems thinking in the international and national context of the nineteenth and twentieth century, has discussed the attractiveness of the concept of the ecosystem for engineering and managerial purposes. Andrew Jamison, "National Political Cultures and the Exchange of Knowledge: The Case of Systems Ecology," in *Denationalizing Science: The Contexts of International Scientific Practice*, ed. Elisabeth Crawford, Terry Shinn, and Sverker Sörlin (Dordrecht: Springer Science and Business Media, 2013), 197ff.

33. For a comprehensive history of early big science in physics and engineering projects, see Peter Galison and Bruce Hevly, eds., *Big Science: The Growth of Large-Scale Research* (Stanford: Stanford University Press, 1992). A more recent discussion of the characteristics of different stages and phases in postwar big science can be found in Olof Hallonsten, *Big Science Transformed: Science, Politics and Organization in Europe and the United States* (London: Palgrave Macmillan 2016).

34. For a study on the uptake of big science thinking in the life sciences, see David C. Coleman, *Big Ecology: The Emergence of Ecosystem Science* (Berkeley: University of California Press, 2010). For the role of large and nationally funded ecological research laboratories in fostering disciplinary solidification during the 1950s and 1960s, see Stephen Bocking, *Ecologists and Environmental Politics: A History of Contemporary Ecology* (New Haven: Yale University Press, 1997).

35. Chunglin Kwa, "Representations of Nature Mediating between Ecology and Science Policy: The Case of the International Biological Programme," *Social Studies of Science* 17, no. 3 (1987).

36. Paul N. Edwards, *A Vast Machine: Computer Models, Climate Data, and the Politics of Global Warming* (Boston: MIT Press, 2010); Agatha C. Hughes and Thomas P. Hughes, *Systems, Experts, and Computers: The Systems Approach in Management and Engineering, World War II and After* (Boston: MIT Press, 2011).

37. For a detailed analysis of the rise and fall of cybernetics and information theory, as well as their remaining relevance for what we now call the information age, see Ronald R. Kline, *The Cybernetics Moment: Or Why We Call Our Age the Information Age* (Baltimore: Johns Hopkins University Press, 2015).

38. Fernando Elichirigoity, *Planet Management: Limits to Growth, Computer Simulation, and the Emergence of Global Spaces* (Chicago: Northwestern University Press, 1999), 32; Henny J. Van der Windt, *En dan, wat is natuur nog in dit land? Natuurbescherming in Nederland 1880–1990* (The Hague: Boom, 1995).

39. Discussing the imagined visions of the planet, as well as the difficulties of early planetary photography, Robert Poole shows what significance imageries can have on our world view. Robert Poole, *Earthrise: How Man First Saw the Earth* (New Haven: Yale University Press, 2008).

40. Jeremy Rifkin, *Biosphere Politics: A New Consciousness for a New Century* (New York: Crown, 1991); Iris Schröder and Sabine Höhler, *Welt-Räume: Geschichte, Geographie und Globalisierung seit 1900* (Frankfurt am Main: Campus, 2005).

41. Höhler, *Spaceship Earth in the Environmental Age, 1960–1990*; Riley E. Dunlap and Angela G. Mertig, *American Environmentalism: The U.S. Environmental Movement, 1970–1990* (New York: Taylor & Francis, 2014); Isenberg, *The Oxford Handbook of Environmental History*.

42. Wolfram Kaiser and Jan-Henrik Meyer, eds., *International Organizations & Environmental Protection: Conservation and Globalization in the Twentieth Century* (New York: Berghahn Books, 2017).

43. Holdgate, *The Green Web*.

44. Glenda Sluga, "UNESCO and the (One) World of Julian Huxley," *Journal of World History* 21, no. 3 (2010); Bharat H. Desai, "UNEP: A Global Environmental Authority," *Environmental Policy and Law* 36, no. 3–4 (2006); also see Natarajan Ishwaran, Ana Persic, and Nguyen H. Tri, "Concept and Practice: The Case of UNESCO Biosphere Reserves," *International Journal of Environment and Sustainable Development* 7, no. 2 (2008); Chloé Maurel, "L'UNESCO, un Pionnier de l'Ecologie?," *Monde(s)* 1, no. 3 (2013); James P. Sewell, *UNESCO and World Politics: Engaging in International Relations* (Princeton: Princeton University Press, 2015).

45. See Alexis Schwarzenbach, *Saving the World's Wildlife: WWF—the First 50 Years* (London: Profile Books, 2011). Similarly, the political philosopher Robert Lampert has studied the first twenty-five years of Friends of the Earth International (founded in 1971), while the environmentalist Rex Weyler has published an insightful history of Greenpeace. Robert Lamb, *Promising the Earth* (London: Routledge, 2012); Weyler, *Greenpeace: How a Group of Ecologists, Journalists, and Visionaries Changed the World*.

46. Thomas Robertson, "'This Is the American Earth': American Empire, the Cold War, and American Environmentalism," *Diplomatic History* 32, no. 4 (2008); Lisa M. Brady, "Life in the DMZ: Turning a Diplomatic Failure into an Environmental Success," *Diplomatic History* 32, no. 4 (2008).

47. John W. Meyer et al., "The Structuring of a World Environmental Regime, 1870–1990," *International Organization* 51, no. 4 (1997); Jacob D. Hamblin, "Gods and Devils in the Details: Marine Pollution, Radioactive Waste, and an Environmental Regime circa 1972," *Diplomatic History* 32, no. 4 (2008): 540; John R. McNeill and Corinna R. Unger, eds., *Environmental Histories of the Cold War* (Cambridge: Cambridge University Press, 2010).

48. John McCormick, "The Origins of the World Conservation Strategy," *Environmental Review* 10, no. 3 (1986); Roderick P. Neumann, "The Postwar Conservation Boom

in British Colonial Africa," *Environmental History* 7, no. 1 (2002). Also see Paul R. Greenough and Anna L. Tsing, *Nature in the Global South: Environmental Projects in South and Southeast Asia* (Durham, NC: Duke University Press, 2003); Dan Brockington, *Fortress Conservation: The Preservation of the Mkomazi Game Reserve, Tanzania* (Bloomington: Indiana University Press, 2002). In general, this tension between global and local ways of perceiving environmental risk or of planning environmental strategies has recently attained more attention, for instance in the work of the American science studies scholars Sheila Jasanoff and Marybeth Long Martello: Sheila Jasanoff and Marybeth Long Martello, eds., *Earthly Politics: Local and Global in Environmental Governance* (Cambridge, MA: MIT Press, 2004).

49. Stephen Macekura, *Of Limits and Growth. The Rise of Global Sustainable Development in the Twentieth Century* (Cambridge: Cambridge University Press, 2015).

50. Iris Borowy, *Defining Sustainable Development for Our Common Future: A History of the World Commission on Environment and Development (Brundtland Commission)* (London: Routledge, 2013).

51. Alessandro Antonello, "Protecting the Southern Ocean Ecosystem: The Environmental Protection Agenda of Antarctic Diplomacy and Science," in *International Organizations & Environmental Protection: Conservation and Globalization in the Twentieth Century*, ed. Wolfram Kaiser and Jan-Henrik Meyer (New York: Berghahn Books, 2017). Similarly, policy scholar Peter Haas, in his work on the convention against pollution in the Mediterranean coastal area, has shown how the existence of expert regimes can explain the outcome of past negotiations concerning antipollution policies: Peter M. Haas, *Saving the Mediterranean: The Politics of International Environmental Cooperation* (New York: Columbia University Press, 1990).

52. Important works to mention here include Michael Egan, *Barry Commoner and the Science of Survival: The Remaking of American Environmentalism* (Cambridge, MA: MIT Press, 2007); Jacob D. Hamblin, *Arming Mother Nature: The Birth of Catastrophic Environmentalism* (New York: Oxford University Press, 2013); Wolfram Kaiser and Jan-Henrik Meyer, eds., *International Organizations & Environmental Protection: Conservation and Globalization in the Twentieth Century* (New York: Berghahn Books, 2017). Haas, for example, who has employed the term "hybrid community," has used it only to describe the complementary reliance and division of labor among experts of different disciplines within the epistemic community to understand aspects of the problem at stake. Peter M. Haas, "Banning Chlorofluorocarbons: Epistemic Community Efforts to Protect Stratospheric Ozone," *International Organizations* 46, no. 1 (1992): 187–224.

53. E.g., Dorothy Nelkin, *Nuclear Power and Its Critics: The Cayuga Lake Controversy* (New York: Cornell University Press, 1971); *Controversy: Politics of Technical Decisions* (New York: Sage Publications, 1979).

54. E.g., Harry M. Collins, *Changing Order: Replication and Induction in Scientific Practice* (Chicago: University of Chicago Press, 1992).

55. Sheila Jasanoff, "Genealogies of STS," *Social Studies of Science* 42, no. 3 (2002): 339–40; Sismondo, "Science and Technology Studies and an Engaged Program," 14–15.

56. Ronald N. Giere, "Controversies Involving Science and Technology: A Theoretical Perspective," in *Scientific Controversies: Case Studies in the Resolution and Closure of Dispute in Science and Technology*, ed. Hugo Tristram Egelhardt and Arthur L. Caplan (Cambridge: Cambridge University Press, 1987), 125–50.

57. In this book, I study expertise as something relational that is assigned by others, in this case policymakers and diplomats, who recognize the experts' authority. The claiming of expertise by disassociation from lay knowledge has been much discussed since Thomas F. Gieryn formulated the concept of boundary work in: "Boundary-Work and the Demarcation of Science from Non-Science: Strains and Interests in Professional Ideologies of Scientists," *American Sociological Review* 48, no. 6 (1983). My research for this book, however, is less concerned with the study of the relationship between experts and different lay publics than with experts in policy decisions and international political debates between diplomats.

58. Stephen Hilgartner, *Science on Stage: Expert Advice as Public Drama* (Stanford: Stanford University Press, 2000).

59. Hilgartner has shown how in the 1970s and 1980s the unified and authoritative public voice of American scientists at the National Academy of Sciences kept from the public view hefty debates between committee members on the validity of nutrition claims. However, alongside scientists' disagreements on the correctness of scientific claims, I am equally interested in discussions on scientific policymaking and methods for environmental governance.

60. Wiebe E. Bijker, Roland Bal, and Ruud Hendriks, *The Paradox of Scientific Authority: The Role of Scientific Advice in Democracies* (Cambridge, MA: MIT Press, 2009), 30; Hilgartner, *Science on Stage*, 4; Bruno Latour, *We Have Never Been Modern* (Cambridge, MA: Harvard University Press, 1993).

61. Jasanoff, *States of Knowledge*; Miller, "Climate Science and the Making of a Global Political Order."

62. Sheila Jasanoff, *The Fifth Branch: Science Advisers as Policymakers* (Cambridge, MA: Harvard University Press, 1990), 247–50.

63. Roger A. Pielke, *The Honest Broker: Making Sense of Science in Policy and Politics* (Cambridge: Cambridge University Press, 2007), 14.

64. Evert Peeters, Joris Vandendriessche, and Kaat Wils, eds., *Scientists' Expertise as Performance: Between State and Society, 1860–1960* (London: Routledge, 2015).

65. For the self-fashioning of experts, see Hilgartner, *Science on Stage*. For the construction of expert roles, see Sheila Jasanoff, *Science at the Bar: Law, Science, and Technology in America* (Cambridge, MA: Harvard University Press, 2009); Arie Rip, "Constructing Expertise: In a Third Wave of Science Studies?," *Social Studies of Science* 33, no. 3 (2003).

66. During the course of this project, I had the chance to talk extensively to several historical actors about their firsthand experiences: Gina Douglas (18 and 19 October 2013) and Jennifer Norman (8 March 2014), who both worked closely with Max Nicholson during the IBP and who met with me in person, and IUCN veterans Lee Talbot and Duncan Poore, who responded to my questions in written form (Talbot: 6 and 8 July 2016; Poore, through Jennifer Norman: 13 March 2014).

 CHAPTER 1

Old Hands, Pastures New

IUCN and the New Environmental Age

Introduction

In the 1960s, the beginning of what has been called "the environmental age," the members of the International Union for Conservation of Nature and Natural Resources (IUCN) were old hands in the conservation game.[1] Providing an official platform for naturalists and scientists interested in international nature protection for over a decade, the origins of their sprawling network dated back to the early 1900s.[2] During the 1960s and 1970s, however, the members of this established network were moving into the new pastures of international environmentalism and environmental politics. In fact, the two decades produced radical changes in the scientific and political contexts in which IUCN was situated. The period saw the rise of a new environmental discourse on the limits of the planet's carrying capacity, as well as the rise of international environmental politics and conventions.[3] In the postwar decades, advances in medical research, in the area of antibiotics for example, as well as agricultural innovations linked to the Green Revolution, contributed to the rapid growth in world population numbers, from 2.5 billion in 1950 to 4 billion by 1970.[4] With this came the growth of cities and suburban areas, which relied heavily on the use of oil, coal, and other nonliving and living natural resources. While air and water pollution intensified, forests and wildlife populations rapidly declined. In this respect, historians have proposed that the swell of new environmentalism that emerged in the 1960s was a direct reaction to this wave of postwar economic growth, globalization, and a new planetary concern over environmental degradation.[5]

Historians of environmental politics and international organizations have often represented the new environmental age—some speak of an environmental revolution—as a break with the exploitative thinking of previous decades. This included a growing public environmental awareness and the emergence of new environmental groups and policy instruments to shield forests from acid rain, schools of fish from overexploitation, and tigers from the destruction of the rainforest.[6] Two prominent examples of relatively early national and international environmental institutionalization are the ratification of the United States National Environmental Policy Act (NEPA) of 1970 and the foundation of the United Nations Environmental Program (UNEP), a bureau established in 1972 to coordinate the environment-related efforts of the United Nations' (UN) special agencies and member governments. The fact that the period saw an unprecedented number of international conventions for environmental protection has found much attention in the literature.[7] These celebrated achievements of environmentalism were paralleled by important changes in global politics. Postwar international relations of decolonization and the Cold War radically changed the geopolitical power distribution within the global order. These new political fault lines influenced the transnational negotiations on the global environment, while the management and distribution of natural resources played an important role in international relations in novel and significant ways.

In addition to accounts that present the 1960s and 1970s as a period of vast and sweeping changes, scholars writing on the history of nature conservation have often pointed to the long tradition of naturalists concerned with the global environment.[8] Conservationists in IUCN's own records tend to stress the continuous relevance of their work and of IUCN as an organization throughout the twentieth century. In particular, these authors from nature protection organizations have pointed to conservationists' long-standing engagement in international environmental negotiations and the establishment of environmental projects and protected areas around the world.[9] Reflecting on notions of both rupture and continuity, this chapter aims at a better understanding of what the intellectual and political changes associated with the environmental age of the 1960s and 1970s meant for discussions on environmental expertise and the role of nature conservation advice therein. It does so by looking at how the new environmental age was met by established scientists in IUCN, which at the time was the oldest and largest international conservation organization concerned with broadly defined environmental protection. In particular, this chapter is concerned with the kinds of opportunities and challenges IUCN conservationists faced in negotiating their position and their expert authority.

Conservationists affiliated with IUCN welcomed the new environmental interest of the 1960s, seeing it as a confirmation of their long-standing agenda.[10] Despite the emergence of a wide range of new alternative groups, for active IUCN members the authority in environmental questions firmly lay with politically neutral scientific institutions like their own. Their long engagement in science-based and broadly defined conservation helped them adapt to the new, broad environmental discourse by linking the management and the conservation of natural resources in an overarching framework of ecosystem ecology. As self-proclaimed neutral scientific advisors, IUCN members were open to cooperation with experts in various disciplines, offering their ecological know-how to actors from the worlds of development, agriculture, and industry. IUCN's network, which had its origins in early twentieth-century conservation societies, and which had been firmly established in the early post-war spirit of scientific universalism, helped the Union's members to present their organization as a platform that could circumvent Cold War tensions between regional camps. Similarly, in the age of decolonization, their experience with nature conservation in diverse geographical regions made them, in their own eyes, legitimate scientific mediators between local and transnational interests.

The new era, however, also came with new challenges, which leading IUCN conservationists mastered only partially. IUCN's scientific authority did not remain unquestioned when new scientific groups with radically different worldviews offered alternatives to ecosystem ecology as the scientific underpinning to environmental problem-solving. Moreover, in the charged diplomatic negotiations of postwar international relations, IUCN elites' claims to scientific neutrality were never fully recognized. Often IUCN conservationists' claims of impartiality and scientific internationalism were rooted in traditional ideas on the sublimity of scientific knowledge over politics, rather than in forms of international representation. These ideas were at odds with the intergovernmental ideology of postwar organizations such as the UN, with which conservationists at IUCN were trying to liaise. Within IUCN, scientists from the global North remained dominant. As a consequence, the network struggled with the politics of globalization and internationalization of the 1970s.

A Long Tradition of Organized Nature Protection

Before turning to the ways in which conservationists affiliated with IUCN tried to adapt to the new environmentalist context, it is important to explain the position from which they encountered different scientific groups and political ideologies. Traditions of environmentalism and conservation, in the Western sense, reach back as far as the time of modernization, exploration, and industrialization of the eighteenth century. Early ideas on the protection of nature have often been linked to a corresponding geopolitics of natural resources.[11] In the late eighteenth and early nineteenth centuries in British and French colonies, local environmental philosophies blended with new scientific ideas on agriculture and reasonable land use, and were taken back to Europe. Eighteenth-century research in the broad field of natural history expanded earlier traditions of classifying and collecting by adding a new empirical interest in the internal dynamics of natural environments. In this respect, European naturalists and explorers such as Joseph Banks and Alexander von Humboldt brought back not only novel natural specimens but also new ideas on the inner workings of nature, humankind's place in it, and human responsibility toward nature.[12] In the nineteenth and early twentieth centuries, it was also in the colonies that the first cross-national institutional and legal instruments created to protect the environment emerged, noticeably from fishery, forestry, and hunting regulations on local resources and wildlife. In the nineteenth century, an intricate web of environmental laws had been adopted on the fringes of European empires, such as French Mauritius and the British Cape Colony. Similarly, the British Society for the Preservation of the Fauna of the Empire, established in 1903 by British naturalists in Africa, is an example of an early conservation institution with colonial roots.[13]

During the late nineteenth century, another strong environmental tradition emerged in the Americas. Environmental historians have especially highlighted the strong spiritual dimension of North American environmentalism, as well as the glorification of wilderness and the sublime that underlay America's early conservation movement at the turn of the last century. The works of the nineteenth-century American naturalists George Perkins Marsh and Henry Thoreau have become iconic examples of this type of early American conservationism.[14] At the end of the nineteenth century, their writings inspired naturalists in the United States to form two early and influential conservation organizations, the Sierra Club and the Audubon Society.[15] Yet, besides spiritual deference, economic interests also played an important role in the American conservation movement. Around 1900, one of the earliest legal agreements between Canada and the United States, which would now qualify as environmental, pertained to the hunting of fur seals.[16]

Looking at the early stages of nature conservation efforts, one finds both the spiritual experience of unity in nature and an economic interest in the extraction and development of living natural resources. International cooperation was also high on the agenda. Around 1900, naturalists on the British Isles and the European continent came together to form the first organizations for nature protection. These included the British Commons Preservation Society of 1899, the Society for Preservation of Nature Monuments (Vereniging tot Behoud van Natuurmonumenten) founded in 1905 in the Netherlands, and the Commission for the Protection of Nature Monuments and Prehistoric Sites of the Swiss Natural History Society of 1906, among others.[17] Members of these regional committees began to reach out to each other by organizing the first international conferences concerned with the protection of nature. It was at these conferences that naturalists and conservationists from Europe and the United States began to think of organizing themselves beyond existing national and regional forums. In fact, as the historian Corey Ross has explained, conservation was part of an early scientific internationalism, and, around 1900, nature conservation was "one of the key realms" of transimperial and international scientific cooperation.[18] In 1909 at a conference in Berne, the Swiss naturalist Paul Sarasin, president of the newly founded Swiss League for the Protection of Nature, suggested that an international union should be formed. By 1913, sixteen European countries, the Russian Empire, and the United States agreed on the *Act of Foundation of a Consultative Commission for the International Protection of Nature*. Yet, only one year later, these international plans were dropped when many of the delegates' home countries found themselves at war.[19]

Soon after the end of World War I, international deliberations and meetings resumed. In 1922, a first international success for nature protection was marked at an informal meeting in London, which resulted in the establishment of the International Council for Bird Preservation (ICBP) by representatives of the North American Audubon Society, the British Royal Society for the Protection of Birds, and the French equivalent, the Ligue pour la Protection des Oiseaux. In 1923 at another international congress, this time in Paris, the Dutch naturalist Pieter van Tienhoven convinced delegates from France, Belgium, and the Netherlands to form a Committee for the Protection of Nature to be integrated into the newly formed League of Nations.[20] While Tienhoven's attempts remained unsuccessful, in 1928 the same countries set up the Office Internationale de Documentation et de Corrélation pour la Protection de la Nature. It took until 1934, however, before an International Office for the Protection of Nature (IOPN) was founded. Seated in Brussels, the IOPN was the first conservation organization that was broad in scope and ambitious enough to tie together naturalist organizations on both sides of the Atlantic.[21] However, soon IOPN projects were disrupted by yet another war.

In the mid-1940s, a joint venture by the former IOPN, the Ligue Suisse, and the ICBP eventually succeeded—in the spirit of postwar conferences, cooperation, and peace initiatives—in infusing internationally organized nature protection with new momentum. In 1948 in the French city of Fontainebleau, the International Union for the Protection of Nature (IUPN) was founded on the initiative of former IOPN members.[22] Building on various national and regional conservation traditions, the founding members of IUPN drafted a twofold program that incorporated the conservation of wild and threatened species, as well as broader ideas on the management and use of living and nonliving natural resources. IUPN functioned as an international platform and advisory body for many nationally and regionally organized groups of naturalists, often concerned with different aspects of the protection of individual species, landscapes, or habitats. Two of IUPN's commissions, the Commission on National Parks and the Species Survival Commission (SSC), especially retained strong links with affluent North American and British organizations. Although individual IUPN members were involved in local conservation projects on the ground, the main role of IUPN and its commissions was a coordinating and advising one, collecting information on the state of the environment and providing conservation guidelines to its different members. In the late 1940s, IUPN had started to collect data on endangered wildlife and habitats, and, by the 1960s, the *Red List of Threatened Species* by the SSC and the *United Nations List of National Parks* constituted focal points in the Union's program.[23] While nature conservation in the so-called developed world was well established, the decolonization of many former colonial regions, believed to be rich in relatively undisturbed natural areas, directed the attention of many IUPN conservationists to the fauna and flora of the global South. Projects such as the SSC's expeditions to the Middle East, Africa, and Asia were part of these initiatives.[24] As a consequence, much of the Union's efforts leading up to the early 1960s focused on maintaining formerly colonial conservation areas on the African and Asian continents, which were home to big game and other rare or unique species of mammals and birds.[25]

In parallel, however, the members of IUPN also looked to broader discussions on the sustainable and responsible use of natural resources in different geographical regions. The topic gained additional traction in the postwar decades through the works of influential thinkers, such as Julian Huxley, chairman of the British Committee for the Conservation of Wildlife, first director general of UNESCO, and one of the Union's founding fathers. In the 1920s and 1930s, Huxley was part of a particular school of politically interested Oxford biologists. The group's ideas on ecology promoted an outlook on human society in which every generation was responsible for maintaining the balance between people and nature for future generations.[26] In the interwar period, Huxley had presented these ideas to the League of Nations. In 1949, it was again at Huxley's instigation that UNESCO convened the Technical Conference in Lake Success, New York, to discuss the future program of IUPN. The conference was a follow-up of the previously held UN Scientific Conference on the Conservation and Utilization of Natural Resources, which had been concerned with the problematic nature of international development programs and the need for a reasonable utilization and careful exploitation of natural resources—themes that were already at the heart of several UN agencies, especially FAO and UNESCO.[27] Huxley lobbied successfully for the inclusion of resource management into IUPN's program. A similar, holistic understanding of nature protection, including the preservation of threatened species, landscapes, and habitats, as well as the management of natural resources, was further emphasized by IUPN's name change to the International Union for Conservation of Nature and Natural Resources in 1956, which proclaimed a broader base for conservation.[28]

While IUCN members were thus concerned with topics ranging from national parks to local development projects, what held them together was a shared reliance on the scientific study of nature. Early on, IUPN's rhetoric focus had been on ecosystem ecology, at the time a rather young field of scientific inquiry, as the intellectual underpinning for their conservation advice.[29] In the 1930s and 1940s, ecosystem ecology had become known through the work of the British botanist Arthur Tansley on the interactions between organisms and their environment. Tansley defined the ecosystem as "a particular category [of] physical systems," containing both organisms and inorganic components in a "relatively stable equilibrium" and existing in "various kinds and sizes."[30] While Tansley's definition was most widely recognized among leading European conservationists, in Russia the biochemist Vladimir Nikolayevich Sukachev coined a similarly influential concept, *biocenosis*, describing interdependent, local systems of organisms and their environment that he believed existed everywhere in nature. Although in the 1930s and 1940s there was no agreement among the members of the scientific community on whether ecosystems presented stable entities in nature, or whether the concept merely constituted an intellectual construct to study natural processes, the ecosystem soon came to be seen as a useful tool to examine and describe the interaction between species, populations, and their habitat. In the 1940s, ecologists were using the concept of the ecosystem to study the links between plant and animal communities in exchange with their nonliving environment.[31] By the mid-1950s, influenced by early computer technology and cybernetics, a new holistic approach to ecosystem ecology emerged. From the 1950s onward, the work of leading ecologists, such as the British biologist G. Evelyn Hutchinson and his American colleague Robert McArthur, focused on the different functions of different organisms and types of environments within ecosystems, and the role of ecological niches, especially. The beginning of modern ecology was marked in the 1960s by the work of two American brothers, Eugene and Howard Odum, who studied and modeled energy flows within closed ecological systems of various scales.[32] Based on the research of the Odum brothers, in the 1970s and 1980s, Heinz Ellenberg, John Ovington, and Paul Duvigneaud began to work with the concept of biomes—larger ecological regions—each containing a particular type of ecosystem.[33]

The origins of global conservation schemes examined in this book thus coincided with important episodes in the history of ecosystem ecology. Ecosystem ecology, in turn, became the scientific foundation for the conservation work that the protagonists of this book proposed and pursued. Institutionally, IUPN's turn to ecology was manifested in 1954 by the creation of the Commission on Ecology, which employed the Union's first staff ecologist, the American Lee Talbot (figure 1.1).[34] Ecology, so it seemed, offered ways to reconcile different traditions in nature conservation. It appeared to provide the scientific foundation both for the protection of threatened species in particular regions and a broader type of conservation advice in reference to resource management around the globe. It was for its appropriability that ecology was actively embraced.

In the early 1960s, the entrenchment of broader environmental concerns combined with a global scope of the science of conservation was in fact crucial

Figure 1.1. Lee and Martie Talbot raiding a lioness in Kenya, 1960. Reproduced from HUA HJC: HUG (FP) 78.75p, Folder 2, with kind permission of the Harvard University Archives.

Conservation defined as resource management could involve the active balancing of population sizes. Such field experience in large conservation areas, often located in the global South, was an important aspect of conservation expertise.

to the self-identification of IUCN conservationists. At IUCN's Seventh General Assembly of 1960, the "application of ecological knowledge in technical assistance and resource development programmes throughout the world" was presented as one of the Union's main objectives.[35] Moreover, based on their broad ecological knowledge, IUCN scientists aimed at an international advisory function at the United Nations (UN), and demanded a say in how their advice was to be implemented. At the Eighth General Assembly in 1963, the actual and potential role of IUCN as an environmental organization was redefined as not only informing regional and national governments, but "encouraging and coordinating the conservation efforts in all parts of the world."[36] In this, IUCN elites based their self-identification on their uniquely scientific and holistic approach. After all, IUCN, like "no other institution," was concerned with the natural world and all natural resources, IUCN President François Bourlière proudly proclaimed in his address to the Ninth General Assembly in 1966.[37] With this broad and ambitious interpretation of ecology-based and globally relevant conservation, IUCN and its members entered the 1960s, welcoming a more widespread environmental concern as an opportunity to promote their cause and to prove their expertise.

The New Environmentalism of the 1960s and 1970s

New to the environmental age of the 1960s and 1970s was the broadly shared understanding that problems pertaining to the natural environment and to society's dependence on natural resources could not be discussed separately.[38] In several ways, the concept of the ecosystem had taken root in society. In 1969, Mohammad Taghi Farvar and John Milton, two conservationists at the American Nature Conservancy, one of the oldest IUCN member organizations, nicely captured this concurrence of the environmental consequences of postwar pollution, population growth, and natural resource depletion, calling it the "unforeseen international ecological boomerang."[39] By the early 1970s, concerns for environmental and societal threats were closely interwoven in the mind of the public and in the arguments of leading intellectuals. Rarely before had so many scientists spoken out publicly about highly political topics such as population growth and the finiteness of natural resources.[40] For many opposition members, public intellectuals, and peace activists, the abhorrence of humankind's destruction of their own planet went hand in hand with antimilitary sentiments and calls for social reform. This was most visible in the bestselling publications of two American biologists, Paul R. Ehrlich's *The Population Bomb* from 1968 and Barry Commoner's *The Closing Circle* from 1971. Both authors discussed the negative side effects of modern production on environmental pollution and resource depletion. Yet, whereas the leftist Commoner focused his criticism mainly on the ruthless and exploitative economies of the postwar era, Ehrlich's even more controversial book blamed uncontrolled population growth in the less-developed regions of the global South.[41]

Despite their differences, however, both Commoner and Ehrlich treated environmental degradation and societal development as two sides of the same coin. This emphasis on dependencies—linking questions on pollution, natural resource depletion, human population growth, and economic development—corresponded to an increasingly widespread intellectual sentiment. IUCN conservationists, too, published on the physical limits of the environment. Early in 1972, the British environmental journal the *Ecologist* had published "A Blueprint for Survival," another report demanding social change to prevent the breakdown of the planet's "life-support system."[42] The report had been signed by over thirty leading scientists, including many well-known conservationists with ties to IUCN, such as Huxley, Frank Fraser Darling, and Peter Scott.

In the perception of leading conservationists at IUCN, the use of natural resources and environmental safeguarding were neither contradictory interests
nor beyond their own expertise.

In the period leading up to the new environmental age, several developments in the natural and life sciences had contributed to the emergence of a
new global environmental consciousness. IUCN conservationists got involved
early in these developments. In 1957 and 1958, the International Geophysical
Year of the International Council of Scientific Unions (ICSU), which involved
engaged scientists from sixty-seven countries, marked the first attempt at the
development of transnational and nonmilitary big science.[43] This effort had
been taken up by other large-scale environmental monitoring programs, such
as the World Meteorological Organization's (WMO) World Weather Watch in
1963, and later in 1967, the Global Atmospheric Research Program that scientists at ICSU and WMO ran jointly.[44] Such programs contributed to transnational scientific cooperation, a growing body of data on the planet's physical
environment, and the establishment of international centers where such data
was stored and analyzed. In the early 1960s, similar plans emerged for an International Biological Program (IBP) to investigate and assess the productivity
of ecosystems around the world. From the beginning, leading IUCN conservationists were interested in contributing to the program based on their expertise in ecosystem ecology and environmental management.

In 1964, the English IUCN conservationist Max Nicholson, head of the
British Nature Conservancy, took up the role of convener of the IBP Section
on Conservation of Terrestrial Communities (IBP/CT). In this position, Nicholson began to promote conservation as part of ecological research that would
help explain natural ecosystems, the limits to their exploitation, and therefore
the parameters for the manageability of natural resources. For him, social and
natural systems were closely interrelated, and so were the protection and use of
natural environments. Nicholson, a close friend of Huxley's and fellow founding father of IUPN, was a particularly influential promoter of a broad understanding of conservation (figure 1.2). Although not a biologist by training,
Nicholson was a keen naturalist and bird-watcher and in the 1920s and 1930s
counted many ecologists among his Oxford friends.[45] Nicholson in particular
took much inspiration from the ecosystem research conducted by the Odum
brothers, who understood ecosystems as closed systems of energy flows, which
could be studied and subsequently managed.[46] For the IBP, Nicholson adapted
the ecosystem thinking of the Odums and applied it to societal issues of human ecology, population, and the production of resources in line with Huxley.
This kind of ecosystem ecology, relevant for social questions and calculable,
received a significant push within conservation circles through the IBP.[47]

The idea that the health of ecosystems was biologically determined by the
intactness of energy flows and nutritious cycles provided the foundation for

Figure 1.2. Max Nicholson (Director of the Nature Conservancy) and Jean G. Baer (IUCN President 1958–63) during the Eighth General Assembly, held in Nairobi, Kenya, between 16 and 24 September 1963 (photo ID 2033) © IUCN.

a new global conception of the environment, which was quickly taken up by the members of IUCN and other international organizations: the idea of the global *biosphere*. The concept of the biosphere had originally been coined in the late nineteenth century by the Austrian geologist Eduard Sueß to describe the livable spaces on the Earth's continents. In 1910, it had been reconceptualized by the Russian biologist Vladimir Ivanovich Vernadsky as the self-contained sphere that enabled life on Earth.[48] The concept of the biosphere gained a wider popularity only in the late 1960s, however, when it was re-discovered by Western ecologists such as Hutchinson, who drew heavily on Verdansky's ideas in his own work.[49] In 1968, inspired by this kind of research, UNESCO held an Intergovernmental Conference of Experts on the Scientific Basis for Rational Use and Conservation of the Resources of the Biosphere, better known as the Paris Biosphere Conference.[50] IUCN conservationists as well were early in adopting the biosphere concept. Many IUCN members, several with strong links to UNESCO, participated in the conference, promoting a broad understanding of conservation that was rooted in ecosystem ecology as promoted by Nicholson.

IUCN elites' success in appropriating the concept of the biosphere was demonstrated when one of the major contributors to the Biosphere Confer-

ence, IUCN's director general Gerardo Budowski, was invited to give a keynote lecture at the First International Congress for Ecology held in 1971 in
Finland, titled "The Biosphere to Come."[51] The German-Venezuelan agronomist and forestry specialist Budowski had been involved in Nicholson's IBP/
CT, possessed an impressive scientific track record, and had demonstrated
his leadership skills during his previous employment at UNESCO (figure
1.3).[52] In the lecture, Budowski described the biosphere as a "single super-
ecosystem," encompassing all living organisms on the planet.[53] The manipulation of one factor was apt to "affect the whole system."[54] Therefore, it was
necessary to maintain the ecological relationships within the system as a first
priority. The focus of international environmental measures was thus to be on
a balanced and closed system of utilization, including extraction and conservation. The maintenance of the biosphere's balance was believed to have an
automatic positive effect on all social and economic affairs.[55] At least within
IUCN and UNESCO, this line of argumentation became part of established
thinking.

After a successful participation in the IBP, leading IUCN conservationists
such as Budowski became more ambitious, promoting their competencies

Figure 1.3. Dr. Gerardo Budowski (*left*) and the American ecologist Thane Riney
(*right*) at IUCN, General Assembly, 10th, New Delhi, IN, 24 November–1 December
1969 (photo ID 3967) © IUCN.

in environmental questions beyond the conservation-oriented community alone.[56] Around 1970, it became known that the UN General Assembly, in the preparation of the Second United Nations Development Decade (1971– 1980),[57] was planning an international conference on human-nature relationships. Ideas behind this first UN Conference on the Human Environment, to be held in Stockholm in the summer of 1972, were closely linked in their origins to themes such as international development aid and the maintenance, use, and distribution of natural resources.[58] Budowski played a leading role in steering IUCN's contribution to the Stockholm Conference, where he hoped to assert a strong advisory function for the Union and responsibilities that included the ecological management of natural resources. In his work at Stockholm, Budowski was encouraged and supported by his senior colleague, the renowned American zoologist and IUCN president Harold Jefferson Coolidge (figure 1.4).

Figure 1.4. Harold Jefferson Coolidge, president of IUCN between 1966 and 1972. Reproduced from HUA HJC: HUG (FP) 78.75p, Folder 1, with kind permission of the Harvard University Archives.

Coolidge was a well-connected scientist who had held previous positions at the Harvard Museum of Natural History and the Pacific Science Board of the National Academy of Science. Like Nicholson, Coolidge had been an influential figure during the formative period of IUCN and at the Lake Success meeting of 1949, where he had promoted international cooperation in broadly conceived questions relating to conservation and resource management.[59] In the early 1970s, the Stockholm Conference seemed to provide an excellent opportunity for ecologists such as Coolidge and Budowski to bring IUCN's conservation advice to the attention of politicians, economists, and development experts.

Development agendas had already played a role during the IBP. During the 1960s, Nicholson, for example, had reached out to researchers in social science disciplines, such as the prominent British economist Barbara Ward. In the following years, Nicholson and Ward organized several workshops and scientific gatherings on the themes of economy and ecology. In 1970, they brought together ecologists and development experts at a Columbia University Conference on International Economic Development.[60] In 1971, when Ward founded the International Institute for Environmental Affairs (IIEA), which functioned as an independent policy think tank, Nicholson became a board member.[61] Around the same time, leading IUCN conservationists made contact with development experts at philanthropist institutions, for instance at the North American Aspen Institute of Humanistic Studies, which, like Ward's IIEA, contributed to the UN Stockholm Conference and to dominant development agencies, such as the World Bank. When in the 1970s the World Bank announced that they were planning to hire a first staff ecologist, Nicholson and his peers celebrated this development as a fruit of their labors and a recognition of their ecological expertise.[62]

In the following years, conservationists linked to IUCN not only liaised with development institutes, but also began to address development planners in their publications, dedicating several works to ecologically informed and conservation-minded development advice. A particularly influential work appeared in 1973, titled *Ecological Principles for Economic Development*.[63] The book's main author, the American biologist and zoologist Raymond Dasmann (figure 1.5), was a former UNESCO colleague of Budowski's and held the position of IUCN Senior Ecologist between 1970 and 1977. Next to Nicholson's work, which exemplified the socioeconomic interest of conservation planners of the 1960s and 1970s, Dasmann's *Principles* is another good example of how biological field experience played into particular visions for land development. Already during his student years at Berkeley, Dasmann had participated in local conservation projects and had observed a recurring conflict between the protection of biological areas for research and exploitative industrial interests. Also important was a later episode of field work in Southern Rhodesia, a Brit-

Figure 1.5. *From left to right: standing,* Kenton Miller, Lee Talbot; *sitting,* Ray Dasmann, Gerardo Budowski, David Munro, Duncan Poore, 1987 (photo ID 2052) © IUCN.

ish colony on the territory of modern Zimbabwe, where Dasmann had studied threatened steppe herbivores. Here he investigated how indigenous ways of using and living with wild animals could help in finding sustainable alternatives to exploitive hunting and farming practices.[64] Drawing on these experiences, the small book of *Ecological Principles,* which defined conservation as "the proper management" of natural resources, focused on ecosystems under development pressure.[65] It encapsulated the stance that many IUCN conservationists took during the mid-1970s in questions relating to the development of newly independent countries in the global South.

Despite these attempts to reach out to economists and development planners, IUCN's ecological perspective—broad and inclusive, yet inherently biocentric and based on the idea of inevitable ecological rules—was not readily accepted by alternative intellectual communities. Often these groups had different ideas on environmental governance and the relevance of conservation knowledge. A prominent opponent of IUCN's biologically determined perspective was the American futurist John McHale, connected to the Aspen Institute and advisor to the organizers of the Stockholm Conference. McHale, a former student of the architect and futurist designer Richard Buckminster Fuller, promoted the technoscientific optimism of his mentor, proposing technological and socioeconomic solutions to environmental problems.[66] His visions fell on the sympathetic ears of growth promoters.

Voices such as McHale's became louder when in 1972 the Club of Rome published *The Limits to Growth*, suggesting the unavoidable collapse of the biospheric support system should human population size not decrease.[67] What was controversial about the Club of Rome study was what some perceived as its authoritative tone, in which little space was granted to technological solutions, while a single set of ecological rules was to govern the fate of all the people of the planet, irrespective of their economic situation.[68] At the Stockholm Conference, critical voices rose against the idea of physical limits to the biosphere, coming from UN member states located in the global South, such as China, India, and Brazil. Others, delegates from Chile, for example, demanded financial compensation for the environmental measures that threatened to restrict the industrial advancement of less-developed countries.[69] These controversies over environmental limits negatively affected the uptake of IUCN's ecological guidelines that, in the Union's view, should underlie all forms of land use and development.

Further competition to IUCN's perception of ecological limits and the need to protect the natural workings of ecosystems came from within the life sciences. By the time ecosystem ecology was firmly established in IUCN, more reductionist approaches to biology and agriculture from fields such as molecular biology and genetic engineering attained a new boost and quickly outstripped ecological research in the exaction of funding.[70] Also the UN's Food and Agriculture Organization (FAO), another important player in environment and resources-related topics within the UN, invested more and more effort into genetic sequencing and the conservation of natural resources on a genetic basis, for example in national seed banks.[71] This reductionist method to safeguard and enhance natural resources outside of the ecosystem in which they naturally occurred presented another alternative to the ecosystem approach pursued by leading IUCN conservationists such as Nicholson, Budowski, and Dasmann.

In some ways, conservationists managed to circumvent potential conflicts with other scientific groups by referring to widely shared and uncontroversial concepts, such as the global biosphere, or by building conceptual bridges.[72] In the 1970s, conservationists began to stress the links between ecosystems conservation and the protection of the natural genetic resources these ecosystems contained, promoting theories on the benefits of biological diversity for the stability of natural environments. This line of reasoning can be found in Duncan Poore's *Ecological Guidelines for Development in Tropical Rain Forests*, published in 1976. In this guidebook, IUCN's Poore (figure 1.5) presented ecosystems as containers for the genetic potential of populations and communities.[73] The same broad approach, which seemed to bridge genetics research and ecosystem conservation, also underlay IUCN's contribution to the *World Conservation Strategy* of 1980, an attempt to align the work of different UN agencies and IUCN on the themes of conservation and development.[74]

Despite some successes in linking conservation to the broader environment and development discourse, with the entrance of new actors, negotiations on shared approaches to the protection and management of the natural environment demanded a constant adaption of the ecological arguments that underlay the reasoning of leading IUCN conservationists. But the growing environmental task field was not the only challenge that IUCN members faced with the onset of the new environmental age. In addition to negotiating their expert roles in relation to other scientific groups, conservationists were eager to embed international conservation efforts in the emerging field of international environmental politics.

New Tensions in Postwar International Relations

During the 1960s and 1970s, international politics was arguably split alongside two axes: along the East-West axis of the Cold War, and the growing North-South divide propelled by the diverging interests of industrialized and less-industrialized countries. As long as IUCN conservationists engaged in international research projects, such as the IBP, in which national representation played only a small role, these political tensions affected their work to a lesser degree. In IUCN's own network, which comprised a mix of governmental and nongovernmental members, the ideology of scientific neutrality had been an important cornerstone since IUCN's foundation.[75] In the 1960s, in addition to their broad ecological approach, which allowed contributions to resource planning and conservation questions, the universal validity of ecological knowledge, as well as the need to spread this knowledge to all world regions, was at the core of the IUCN conservationists' professional identity.[76] Yet in their attempts to liaise with UN agencies, such as UNESCO and FAO, Cold War politics and international development diplomacy attained a new relevance for leading conservationists representing the interests and the agenda of the Union, which brought about both new opportunities, but also new challenges.

The two decades studied in this book mainly fall between the détente of East-West relations after the Cuban Missile Crisis of 1962 and the widening rift between the USSR and the U.S. government following the Soviet invasion of Afghanistan in 1979.[77] In some ways, historians have claimed, environmental topics that arguably pertained to all world regions helped diplomats build additional diplomatic bridges across the Iron Curtain.[78] In fact, historians of science, politics, and the environment have described how new surveillance technologies from the Cold War period created a new vision of the global environment. Sabine Höhler, Naomi Oreskes, John Krige, and others have described the ambivalent tension that emerged during this period between the notion of a divided humanity and a single planetary system on which all political camps depended to the same degree.[79]

Despite the rise of a global consciousness, however, Cold War politics continued to complicate the options for international diplomacy throughout the détente period, which spilled over into environmental topics as well.[80] This was especially visible within UNESCO. Although the USSR and several European countries of the Eastern bloc had joined UNESCO in 1953 and 1954, rivalries between East and West had had a negative impact on UNESCO's ability to find unified goals in any of its global agendas. Authors have pointed to instances of internal McCarthyist witch hunts in the 1950s and Soviet protests against the new Western director general, the French philosophy professor René Maheu, in the 1960s.[81] Only in the 1980s did the USSR participation in UNESCO's environmental projects, such as the Man and the Biosphere Program (MAB), gain significant momentum. FAO projects on desertification or productivity had no Soviet participation at all, as the USSR refused to give insights into their agricultural programs.[82] The organization of the UN's Stockholm Conference, too, had suffered from Cold War tensions. Several Eastern countries had decided to boycott the international gathering because the German Democratic Republic (GDR) had not been invited.[83]

In light of these continued tensions between East and West, IUCN elites saw the opportunity to distinguish themselves as a politically neutral, scientific expert platform for questions pertaining to the global environment. In this, IUCN members banked on their official status as a nongovernmental scientific organization (NGO). Despite the fact that the Union's member list included several national governments, in 1966 IUCN's president Bourlière stressed in his address to the Ninth General Assembly that the organization acted "without the political influence of governments" and need "not be burdened by bureaucratic dictates," being "open to all countries."[84] In this statement, Boulière invoked the commitment to scientific internationalism that had marked the Union's early years: one of the first decisions had been that the Union was to function as an "international arbiter."[85]

Throughout the 1960s and 1970s, IUCN elites continued to build on this tradition of self-proclaimed scientific neutrality. In particular, this self-image was built on three propositions. First, IUCN's members postulated that the Union was an "independent international body" and therefore was "without the political influence of governments."[86] Second, IUCN was "free from political prejudices," and this neutrality made IUCN's own work "truly international in character."[87] Third, IUCN's members derived from this that they had full rights to ensure the "perpetuation and enhancement of all the living world" and to make sure that all governments would take into consideration the "ecological value" of the ecosystems on their territory.[88] Being beyond reproach in their own eyes in terms of questions on internationality and political neutrality, leading IUCN conservationists claimed responsibility for the global environment in its entirety. At IUCN, scientists could meet without political barriers. Already in the late 1950s, IUCN's Education Committee was trying to reach out to Eastern European countries.[89] In a similar attempt, conservationists had been zealous to include experts from the USSR and its satellite states during the IBP, efforts that had been appreciated by Eastern scientific communities.[90]

This does not mean that IUCN conservationists did not face any Cold War–related difficulties, however. The majority of IUCN members were from Western countries. Eastern experts faced constraints traveling as, for instance, visa issues could prevent USSR scientists from attending important scientific meetings.[91] Their participation in international conservation conferences was low throughout the period.[92] A look at high-ranking executive and scientific positions at IUCN is quite telling (table 1.1). The representation of the Eastern bloc remained low throughout the period. Other communist countries, such as China, were completely absent. For the period from 1960 to 1981, the table shows the respective percentages for the regional representation based on the total number of positions counted (n).[93] Counted were members of IUCN's executive board, later council, who were elected for six years at every General Assembly and were responsible for the Union's programmatic decisions and the preparation of the General Assembly. The table also includes the chairpersons and vice chairpersons of IUCN's commissions and committees, who occupied the highest programmatic functions and after the 1970s constituted the advisory council to the executive board.[94] Further, the table includes the secretary general, later director general, responsible for the Union's finances and personnel, and the senior ecologist, later scientific director, as the highest scientific staff. Technical positions, such as commission chairpersons, were especially dominated by Western European scientists.

Nevertheless, the proclaimed potential of IUCN conservationists to function as neutral mediators for scientists and politicians discussing environmental problems was at least partly recognized. In fact, the historical records show that intergovernmental and governmental institutions sought to outsource environmental East-West relations to IUCN. In the mid-1950s, UNESCO officials explicitly indicated that as a nongovernmental organization, IUCN was permitted to have "sheep and goats mingle for the common good."[95] Powerful political parties, too, accepted IUCN as an alternative channel for discussing the environment. At IUCN's General Assembly in 1963, U.S. Secretary of Interior Steward Udall stressed this ideology-neutral sphere of IUCN, calling it a "market-place where men of all nations may share knowledge gained in dealing with resource problems."[96]

Table 1.1 Regional representation in high-ranking administrative and scientific positions within IUCN.

	1960	1963	1966	1969	1972	1975	1978	1981
Africa	9%	9.5%	8%	3%	6%	8%	3%	10%
Australia and Oceania	0%	0%	4%	0%	6%	11%	3%	10%
Central and South America	4%	5%	4%	7%	9%	6%	8%	7%
East Asia	4%	5%	4%	3%	9%	11%	10%	7%
Eastern Europe	4%	5%	0%	7%	0%	3%	8%	2%
North America and Caribbean	22%	14%	16%	19%	20%	22%	34%	27%
USSR	0%	9.5%	8%	9%	15%	8%	5%	5%
West Asia	0%	0%	0%	0%	3%	3%	8%	10%
Western Europe	57%	52%	56%	52%	32%	28%	21%	22%
Eastern bloc	9%	14%	8%	16%	15%	11%	8%	7%
Western bloc	74%	71%	72%	71%	68%	67%	71%	68%
Global South	17%	19%	16%	15%	26%	28%	29%	34%
Global North	83%	81%	84%	85%	74%	72%	71%	66%
n= total number of positions	23	21	25	31	34	36	38	41

Regional categories are listed according to IUCN's own criteria, except for the USSR, which was divided over Eastern Europe and West Asia. The category has been added for the sake of comparison. Furthermore, the regions Eastern bloc, Western bloc, Global South, and Global North have been added as categories of analysis. Counted as belonging to the Global North were countries from Australia and Oceania, Eastern Europe, North America and the Caribbean, the USSR, Western Europe, and Japan. Counted as belonging to the global South were countries from Africa, Central and South America, East Asia (except Japan), and West Asia. Counted as belonging to the Western bloc were countries from North America and the Caribbean, Western Europe (except for neutral Switzerland and Finland), Turkey, and the ASEAN countries. Counted as belonging to the Eastern bloc were satellite states in Eastern Europe, the USSR, Communist China, Vietnam, and Iraq after 1968.

Despite the unequal representation of Eastern and Western scientists within IUCN, the Union served, at least to a certain extent, as a meeting ground for scientists dispatched by different political regimes. In the coming years, several scientists from Eastern bloc countries played an important role in the high executive ranks of IUCN. For instance, the Russian wildlife biologist Andrei Bannikov held the position of vice president of the Union in the 1970s. Together with the Polish Anna Medwecka-Kornas, who had served as intermediary, Bannikov had been involved in the IBP.[97] Another influential figure who further established relations with the USSR was the Russian biologist Lev Shaposhnikov, who chaired the Commission on Education from the 1960s to 1978.[98] Under Shaposhnikov, the commission organized several influential conferences in Africa and Southeast Asia on conservation education from both biological and societal points of view.[99]

Alongside Cold War tensions between East and West, the politics of decolonization also played a major role in negotiating global conservation advice and environmental expertise. Within the UN and its agencies, the decolonization of many African states in the early 1960s had led to a shift in majority from the Northern to the Southern Hemisphere. Such a shift also occurred within UNESCO and FAO, which until the foundation of UNEP in 1972 constituted the main UN agencies concerned with environment-related questions.[100] The UN, in contrast to other big development spenders such as the World Bank, guaranteed Southern countries voting power and the opportunity to meet other leaders from the developing and developed world. In 1971, after a push from Brazil, communist China joined Group 77, a new coalition of developing nations.[101] Together, China and Brazil lobbied for the global South's right to their own industrialization. In December 1971, the representatives of Group 77 brought their demands to the UN General Assembly, pointing to the industrialized nations in the global North as the main cause for current pollution problems and demanding financial assistance to mitigate their own in future.[102]

With these new demands came a challenge to established patterns of aid and assistance coordinated by experts from the global North. Well into the 1960s, development and aid politics had been dominated by the idea that Western science and technology could bring benefits to development countries.[103] The World Bank, founded in 1944, and other development agencies had been built on this idea, and UN Secretary-General U Thant had firmly endorsed the approach. However, toward the end of the 1960s, it became increasingly clear that the export of technology to remote places had not necessarily led to local development. The realization that despite some growth in the 1960s, there was no improvement of the general living conditions in countries of the global South led to new theories of, and strategies for, development. In 1969, this new consciousness was expressed in the World Bank report "Partners in Development," which moved the international focus toward humanitarian issues and promoted the growing integration of demands of the South on the international agenda.[104] The report formed the background to the UN's new development politics of the 1970s, suggesting a restructuring of development and trade structures, as well as of scientific and technological assistance. The emphasis shifted from financial and technical assistance from the global North to the development of local knowledge and small-scale, "appropriate" technologies.[105]

This shift was important, too, for conservationists and their aim to link the environment discourse to UN development politics. IUCN had a long-established presence in the Southern Hemisphere. A large number of the Union's efforts had focused on traditional conservation areas on the African and Asian continents, which were home to many large and "charismatic" animals.[106] However, most conservation projects in these regions carried colonial legacies. An important example of such efforts from the early 1960s was a large symposium on the "conservation of nature and natural resources in modern African states" held in Arusha, Tanganyika—now Tanzania—in 1961, together with FAO, UNESCO, and the former colonial Commission for Technical Co-operation in Africa (CCTA). From this emerged the IUCN African Special Program, IUCN's first major regional research program in the early 1960s, led by Nicholson's IBP colleague Edward Worthington. The program aimed at integrating national parks and wildlife protection in the development planning by African states.[107] Further, projects such as the SSC's expeditions to the Near and Middle East, Africa, and Asia, and the resulting report, *A Look at Threatened Species*, were part of these initiatives.[108]

In the changing climate of postcolonial politics, leading IUCN conservationists tried to leave behind their colonial legacy. In particular, they hoped to strengthen their links to the UN system. Especially in this UN context, however, IUCN elites were under pressure to pay more attention to the kind of organizational politics their network represented. In the following years, development topics in particular regions were placed more prominently on IUCN's agenda. In the mid-1970s, after IUCN's participation in the UN Stockholm Conference, IUCN began to hold workshops based on Dasmann's *Ecological Principles* to discuss the ecological dimensions of development in different parts of the so-called developing world, including two conferences in 1974 in Southeast Asia and Latin America.[109] Even so, if they wanted to keep their reputation as a neutral scientific platform, IUCN elites needed to do something about the Northern bias in their own organization.

In the following years, institutional reforms inside IUCN aimed at bringing in additional members from less-developed countries—for example, by introducing "regional councillors" who would function as ambassadors between IUCN and local governments.[110] Martin Holdgate, who served as IUCN's director general between 1988 and 1994, has pointed out that during the late 1960s and the 1970s, the number of IUCN members from countries of the global South increased to almost equal those of the global North.[111] In fact, looking at Table 1.1, we do see a slow but steady increase in scientists from the global South in high technical and administrative positions. In 1975, at IUCN's Twelfth General Assembly, it was decided that the organization needed to pay more attention to the geographical representativeness of its international membership. Funds were to be made available to invite representatives from the periphery to the assemblies and meetings taking place in European countries or the United States, the seat of most conservation organizations.[112] Members reported that through new associations with UNEP and the WWF, field contacts had already been made in many less-developed countries, also during personal visits by the president, board members, and by different chairpersons.[113]

By the mid-1970s, seeking closer contact with the different UN agencies and bureaus, the members of IUCN put more emphasis on a balanced representation. In 1975, IUCN entered the Ecosystem Conservation Group, an alliance with international organizations such as UNEP, FAO, and UNESCO.[114] In this respect, an important and influential liaison was UNEP's first director general, Maurice Strong. Strong was a self-made Canadian businessman who, born into deprived circumstances and without a university education, had worked his way up to an impressive and international career in the oil business. By getting involved in the Canadian International Development Agency (CIDA), Strong eventually attracted the attention of UN Secretary General U Thant and the UN's Economic and Social Council's (ECOSOC) under-secretary, Philippe De Seynes. In 1971, Strong was promoted to the position of Secretary General of the Stockholm Conference.[115] In 1978, Strong (figure 1.6) attained an executive function at IUCN.[116]

Although with Strong, IUCN managed to secure stronger links with development experts, the integration of interests and scientists from the global South remained a challenge for IUCN. When in 1977 a larger council replaced IUCN's executive board, regional representation was the criterion according to which new members were elected. However, this new measure did little to overthrow established hierarchies. In 1978, a first attempt at electing an African councilor failed, and only in 1981 was the continent again represented in the organization's membership.[117] Table 1.1 shows that the integration of conservationists from the global South in high-ranking administrative and technical positions at IUCN went much slower than the change in general membership reported by Holdgate. Instead of replacing experts from the old powers, new experts were appointed in addition to them. Technical positions, especially, remained dominated by Western European and North American scientists until the end of the period. Nevertheless, leading IUCN conservationists of diverse national backgrounds continued to portray an image of the organization's scientific impartiality. At the opening of IUCN's Fifteenth Gen-

Figure 1.6. Portrait of Mr. Maurice Strong (Canada), appointed to the post of secretary-general of the 1972 United Nations Conference on the Human Environment, 30 March 1971 © UN Photo/Grunbaum.

eral Assembly in Christchurch in 1981, President Mohamed Kassas, professor of botany at Cairo University, stressed that IUCN should never become an intergovernmental organization itself, and maintained that the Union's politically independent scientific work was the members' main responsibility.[118] In this rhetoric, scientific expertise was more important than global membership.

In fact, another look at the distribution of high-ranking IUCN positions (Table 1.1) shows that instead of a shift in East-West and North-South relations, we see a shift within Western representation. After 1975, the previous European dominance declined, while the North American membership rose parallel to that of countries from the global South. After the Stockholm Conference, politicians and intellectuals from the United States, where neo-Malthusian fears of resource shortage and overpopulation were especially pronounced, got more involved in global environmental discussions.[119] New conservation projects of the 1970s in newly independent countries were mainly run by Americans, for instance by the members of the New York Zoological Society, rather than by Europeans. In the early 1970s, IUCN itself was led by the American Coolidge, who closely worked with American conservation societies.[120] At the same time, IUCN elites were building stronger relations with UN agencies and international philanthropic bodies, such as the Ford Foundation or the World Bank, which were to a large degree dominated by Americans.[121] At IUCN, several technical positions were now filled by members of a younger generation of American ecologists and biologists, born in the late 1930s, the 1940s, and the early 1950s, such as Peter Jacobs, Jeffrey McNeely, and Daniel Navid.[122]

Expert communities from the global South, however, still met IUCN with skepticism.[123] In the late 1970s, Ossamma M. El-Tayeb, an Egyptian microbiologist from Cairo University working for UNEP, complained to UNEP's deputy executive director, the Kenyan environmentalist Reuben Olembo, about IUCN's "naïve and emotional" conservation approach, which lacked sound regional development guidelines.[124] In fact, UNEP members were busily building up their own alternative environmental network to the Northern-dominated IUCN. In 1975, the Egyptian microbiologist Mostafa Tolba had replaced Strong as executive director of UNEP. This satisfied those who wanted UNEP to be run by a representative of the global South. In the coming years, UNEP would attract more experts from less-developed countries and explicitly focus on development projects in the Southern Hemisphere.[125]

Conclusion

Often the new environmental age has been treated as a counterreaction to the economic expansion and increased industrialization of the postwar decade. By looking at the established scientific group of IUCN conservationists, this chapter has instead drawn attention to the blend of continuities, reform, and disruption that marked the negotiations of conservation advice and environmental expert roles during the 1960s and 1970s. With the rise of widespread environmental thinking, leading IUCN conservationists were confronted with new intellectual approaches, new expert groups, and new tensions in the arena of international politics. Yet, rather than a complete overhaul of existing ways of approaching the environment and its protection, conservation strategies, scientific approaches and political lines of argumentation were adapted and renegotiated within established groups, such as the IUCN network.

For conservationists affiliated with IUCN, the new environmental age brought with it several balancing acts in both the intellectual and the political realm. Their ecosystem approach allowed IUCN elites to connect conservation to many themes that were related to the environment defined in ever-broader terms. At the same time, their idea that ecological rules applied to ever-wider topic areas caused conflicts with new groups of scientists who promoted alternative approaches that seemingly allowed for unlimited economic growth and development. Some of these were easier to reconcile with IUCN's ecosystem conservation approach than others. IUCN elites faced new political challenges, too. With the environment becoming a topic for international political deliberations, scientific organizations such as IUCN had to address the growing political tensions between East and West, and North and South. In the realm of international politics, IUCN's insistence on science and scientific neutrality was only partly successful. While in a few instances IUCN was accepted as a natural platform to discuss environmental topics across the Iron Curtain, in general, high-ranking technical positions remained dominated by Western Europeans and North Americans. This challenged IUCN's credibility, especially in development-related questions pertaining to natural resource management in the global South, a topic high on IUCN's own agenda.

During these continuous negotiations between established conservationists and new environmental groups, decisions on conservation advice and environmental expert roles emerged as compromises and arrangements in which continuities and disruptions intertwined. Following this line of reasoning, the next three chapters of this book will address three expert controversies, examining in detail the complex processes of negotiation, adaption, and change.

Notes

1. I am borrowing this term from the environmental historian Sabine Höhler, *Spaceship Earth in the Environmental Age, 1960–1990* (London: Pickering & Chatto, 2015). For a similar periodization, see Joel B. Hagen, "Teaching Ecology during the Environmental Age, 1965–1980," *Environmental History* 13, no. 4 (2008). For a comprehensive study into the conceptual origins of the concept of the environment, see Paul Warde, Libby Robin, and Sverker Sörlin, *The Environment: A History of the Idea* (Baltimore: John Hopkins University Press, 2018). Also see, Paul Warde, Libby Robin, and Sverker Sörlin, "Stratigraphy for the Renaissance: Questions of Expertise for 'the Environment' and 'the Anthropocene,'" *The Anthropocene Review* 4, no. 3 (2017).
2. E.g., Robert B. Boardman, *International Organization and the Conservation of Nature* (Bloomington: Indiana University Press, 1981), 63.
3. Höhler, *Spaceship Earth in the Environmental Age*, 3; Rachelle Adam, *Elephant Treaties: The Colonial Legacy of the Biodiversity Crisis* (Lebanon, NH: University Press of New England, 2014).
4. John H. Perkins, *Geopolitics and the Green Revolution: Wheat, Genes, and the Cold War* (Oxford: Oxford University Press, 1997), 118ff.; William M. Adams, *Green Development: Environment and Sustainability in a Developing World* (London: Routledge, 2008), 46.

5. E.g., Mark Rupert, *Ideologies of Globalization: Contending Visions of a New World Order* (London: Routledge, 2012); Thomas Borstelmann, *The 1970s: A New Global History from Civil Rights to Economic Inequality* (Princeton: Princeton University Press, 2012), 175ff.

6. For accounts of the rising environmental awareness, especially among Western publics, see Mikael S. Andersen and J. Duncan Liefferink, *European Environmental Policy: The Pioneers* (Manchester: Manchester University Press, 1997); Robert V. Bartlett, Priya A. Kurian, and Madhu Malik, *International Organizations and Environmental Policy* (Westport: Greenwood Press, 1995); Otis L. Graham, ed., *Environmental Politics and Policy, 1960s–1990s* (University Park: Pennsylvania State University Press, 2010); Thorsten Schulz-Walden, *Anfänge globaler Umweltpolitik: Umweltsicherheit in der internationalen Politik (1969–1975)* (Munich: Oldenburg Verlag, 2013); Riley E. Dunlap and Angela G. Mertig, *American Environmentalism: The U.S. Environmental Movement, 1970–1990* (New York: Taylor & Francis, 2014).

7. Andrew C. Isenberg, *The Oxford Handbook of Environmental History* (Oxford: Oxford University Press, 2014), 122, 536; Graham, *Environmental Politics and Policy*; Adam, *Elephant Treaties*.

8. E.g., see William M. Adams, *Against Extinction: The Story of Conservation* (London: Earthscan; Fauna & Flora International, 2004); Rachelle Adam, *Nature, Colonialism and Conservation Organizations: How International Law Became the Response to the Biodiversity Crisis* (Jerusalem: Hebrew University of Jerusalem, 2012); John Sheail, *Nature Conservation in Britain: The Formative Years* (London: Stationery Office, 1998); Raf De Bont, Simone Schleper, and Hans Schouwenburg, "Conservation Conferences and Expert Networks in the Short 20th Century," *Environment and History* 23, no. 4 (2017); Elizabeth Kolbert, *The Sixth Extinction: An Unnatural History* (New York: Henry Holt and Company, 2014).

9. E.g., Martin W. Holdgate, *The Green Web: A Union for World Conservation* (Gland: IUCN, 1999); Boardman, *International Organization and the Conservation of Nature*; Raymond F. Dasmann and Randall Jarrell, *Raymond F. Dasmann: A Life in Conservation Biology* (Bloomington: Xlibris Corporation, 2000); Michael Lockwood, Graeme Worboys, and Ashish Kothari, *Managing Protected Areas: A Global Guide* (London: Earthscan, 2012).

10. E.g., E. Max Nicholson, *The Environmental Revolution: A Guide for the New Masters of the World* (London: Penguin Books, 1970); *The Big Change: After the Environmental Revolution* (New York: McGraw-Hill, 1973).

11. For two excellent works discussing the geopolitics of natural resources, see Peder Anker, *Imperial Ecology: Environmental Order in the British Empire, 1895–1945* (Cambridge, MA: Harvard University Press, 2001); Richard H. Grove, *Green Imperialism: Colonial Expansion, Tropical Island Edens and the Origins of Environmentalism, 1600–1860* (Cambridge: Cambridge University Press, 1995).

12. E.g., see Andrea Wulf, *The Invention of Nature: Alexander von Humboldt's New World* (London: Knopf Doubleday, 2015); Martin Mulligan and Stuart Hill, *Ecological Pioneers: A Social History of Australian Ecological Thought and Action* (Cambridge: Cambridge University Press, 2001).

13. Adams, *Against Extinction*; Adam, *Elephant Treaties*, 15.

14. William Cronon, *Uncommon Ground: Toward Reinventing Nature* (New York: W.W. Norton & Company, 1995), 71; Roderick Nash, *Wilderness and the American Mind*

(New Haven: Yale University Press, 1965). For two classic works of American environmentalism, see George P. Marsh, *Man and Nature* (Cambridge, MA: Belknap Press, 1864); Henry D. Thoreau, *Walden* (Boston: Houghton, Mifflin and Company, 1882).

15. E.g., John R. McNeill and Erin S. Mauldin, *A Companion to Global Environmental History* (Chichester: John Wiley & Sons, 2014); Lawrence Buell, *The Environmental Imagination: Thoreau, Nature Writing, and the Formation of American Culture* (Cambridge, MA: Harvard University Press, 1996); Carolyn Merchant, *American Environmental History: An Introduction* (New York: Columbia University Press, 2007).

16. Mark V. Barrow, *Nature's Ghosts: Confronting Extinction from the Age of Jefferson to the Age of Ecology* (Chicago: University of Chicago Press, 2009), 140–41.

17. David Evans, *A History of Nature Conservation in Britain* (London: Routledge, 1992), 43; John Sheail, *Nature's Spectacle: The World's First National Parks and Protected Places* (London: Taylor & Francis, 2014), 103.

18. Corey Ross, "Tropical Nature as Global Patrimoine: Imperialism and International Nature Protection in the Early Twentieth Century," *Past & Present* 226, no. 10 (2015).

19. Boardman, *International Organization and the Conservation of Nature*, 29.

20. Anna-Katharina Wöbse, *Weltnaturschutz. Umweltdiplomatie in Völkerbund und Vereinten Nationen 1920–1950* (Frankfurt am Main: Campus Verlag, 2012); "Oil on Troubled Waters? Environmental Diplomacy in the League of Nations," *Diplomatic History* 32, no. 4 (2008).

21. Adams, *Against Extinction*, 45; Raf De Bont, "Borderless Nature: Experts and the Internationalization of Nature Protection, 1890–1940," in *Scientists' Expertise as Performance*, ed. Joris Vandendriessche, Evert Peeters, and Kaat Wils (London: Pickering & Chatto, 2015).

22. Wöbse, *Weltnaturschutz*, 284.

23. Boardman, *International Organization and the Conservation of Nature*, 51.

24. Holdgate, *The Green Web,* 85; Roderick P. Neumann, "The Postwar Conservation Boom in British Colonial Africa," *Environmental History* 7, no. 1 (2002); see also Lee M. Talbot and IUCN, *A Look at Threatened Species: A Report on Some Animals of the Middle East and Southern Asia Which are Threatened with Extermination* (Cambridge: Fauna Preservation Society for the International Union for Conservation of Nature and Natural Resources, 1960).

25. For accounts on the colonial legacy of the European conservation movement, see Adam, *Elephant Treaties*; Helen Tilley, *Africa as a Living Laboratory: Empire, Development, and the Problem of Scientific Knowledge, 1870–1950* (Chicago: University of Chicago Press, 2011); Roderick P. Neumann, *Imposing Wilderness: Struggles over Livelihood and Nature Preservation in Africa* (Berkeley: University of California Press, 1998).

26. E.g., see Julian Huxley, *The Stream of Life* (London: Watts, 1926); Julian Huxley et al., *Scientific Research and Social Needs* (London: Watts & Co., 1934); Arthur G. Tansley, *The Values of Science to Humanity: The Herbert Spencer Lecture, Oxford University, 2 June 1942* (London: G. Allen & Unwin, 1942).

27. UNESCO, ed., *International Technical Conference on the Protection of Nature: Lake Success, 22–29 August 1949; Proceedings and Papers* (Paris: UNESCO, 1950).

28. Holdgate, *The Green Web*, 63.

29. Boardman, *International Organization and the Conservation of Nature*, 43. For an example, see Georges Petit, "Protection de la Nature et Ecologie," in *International Techni-*

cal Conference on the Protection of Nature: Lake Success, 22–29 August 1949; Proceedings and Papers, ed. UNESCO (Paris: UNESCO, 1950), 314.

30. Arthur G. Tansley, "The Use and Abuse of Vegetational Concepts and Terms," *Ecology* 16, no. 3 (1935): 306, 299.

31. Frank B. Golley, *A History of the Ecosystem Concept in Ecology: More Than the Sum of the Parts* (New Haven: Yale University Press, 1993), 61ff.; David G. Raffaelli and Christopher L. Frid, eds., *Ecosystem Ecology: A New Synthesis* (Cambridge: Cambridge University Press, 2010), 5; John Sheail, *Seventy-Five Years in Ecology: The British Ecological Society* (Oxford: Blackwell Scientific, 1987).

32. Robert McIntosh, *The Background of Ecology: Concept and Theory* (Cambridge: Cambridge University Press, 1986), 227; Raffaelli and Frid, *Ecosystem Ecology*, 9; Eugene P. Odum, *Ecology* (New York: Holt, Rinehart and Winston, 1963); Howard T. Odum, *Environment, Power, and Society* (New York: Wiley-Interscience, 1970).

33. There are many excellent works on the origins of ecosystem ecology that need no summary here, but there are a few names that are worth pointing out for their influence within conservation circles. For the history of ecology in Britain, see Hannah Gay, *The Silwood Circle: A History of Ecology and the Making of Scientific Careers in Late Twentieth-Century Britain* (London: Imperial College Press, 2013); John Sheail, *Seventy-Five Years in Ecology: The British Ecological Society* (Oxford: Blackwell Scientific, 1987). For the history of ecosystem ecology in the twentieth century, see Joel B. Hagen, *An Entangled Bank: The Origins of Ecosystem Ecology* (New Brunswick: Rutgers University Press, 1992); Frank B. Golley, *A History of the Ecosystem Concept in Ecology: More Than the Sum of the Parts* (New Haven: Yale University Press, 1993); David C. Coleman, *Big Ecology: The Emergence of Ecosystem Science* (Berkeley: University of California Press, 2010). For the links between ecology and scientific and political environmentalism, see Joachim Radkau, *Die Ära der Ökologie: Eine Weltgeschichte* (Munich: C. H. Beck, 2011); Donald Worster, *Nature's Economy: A History of Ecological Ideas*, Studies in Environment and History (Cambridge: Cambridge University Press, 1985); Brian Walker, *Ecologists and Environmental Politics: A History of Contemporary Ecology* (Chicago: University of Chicago Press, 1998). Moreover, there are a number of excellent accounts of the disciplinary contributions of individual ecologists, such as Peter G. Ayres, *Shaping Ecology: The Life of Arthur Tansley* (Oxford: John Wiley & Sons, 2012); Betty Jean Craige, *Eugene Odum: Ecosystem Ecologist and Environmentalist* (Athens, GA: University of Georgia Press, 2002); David M. Richardson, ed., *Fifty Years of Invasion Ecology: The Legacy of Charles Elton* (Oxford: John Wiley & Sons, 2011); Nancy G. Slack, *G. Evelyn Hutchinson and the Invention of Modern Ecology* (New Haven: Yale University Press, 2010).

34. Holdgate, *The Green Web*, 63.

35. IUCN, ed., *Proceedings of the Seventh General Assembly, Warsaw 1960* (Morges: IUCN, 1960), 94.

36. IUCN, ed., *Proceedings of the Eighth General Assembly, Kenya 1963* (Morges: IUCN, 1964), 77.

37. IUCN, ed., *Proceedings of the Ninth General Assembly, Lucerne 1966* (Morges: IUCN, 1966), 101.

38. An early and influential work to mention in this regard is the American biologist Rachel Carson's *Silent Spring*, which describes the severe environmental effects of pesticide use at the time; William Souder, *On a Farther Shore: The Life and Legacy of Rachel Carson*

(London: Crown Publishers, 2012); Rachel Carson, *Silent Spring* (Boston: Houghton Mifflin Harcourt, 1962).

39. Mohammad Taghi Farvar and John P. Milton, *The Unforeseen International Ecologic Boomerang: Conference on the Ecological Aspects of International Development* (New York: American Museum of Natural History, 1969). The volume presented a preliminary account of a conference in 1969 at Airlie House near Washington, sponsored by the North American Conservation Foundation and the Center for the Biology of Natural Systems at Washington University, Missouri. At the conference, environmental scientists, representatives of international organizations, and conservationists, including IUCN affiliates Dasmann and Talbot, came together to discuss the environmental consequences of international economic and technological development, including development aid from the global North to the global South. The volume presented a prequel to the more extensive proceedings of the conference, published three years later and titled *The Careless Technology*: Mohammad Taghi Farvar and John P. Milton, eds., *The Careless Technology: Ecology and International Development; The Record of the Conference on the Ecological Aspects of International Development, December 8–11, 1968, at Airlie House, Warrenton, Virginia* (Garden City, NY: Natural History Press, 1972).

40. For the links between science and 1970s counterculture, see David Kaiser and W. Patrick McCray, *Groovy Science: Knowledge, Innovation, and American Counterculture* (Chicago: University of Chicago Press, 2016).

41. In 1968, Ehrlich predicted an apocalyptic outcome of a growing human population, global resource shortages, and famine. Three years later, Commoner presented a leftist answer to Ehrlich's doomsday scenario, presenting capitalist technologies as the main cause for environmental degradation. Paul R. Ehrlich, *The Population Bomb* (New York: Buccaneer Books, 1968); Barry Commoner, *The Closing Circle: Man, Nature, and Technology* (New York: Knopf, 1971).

42. Frank B. Goldsmith, "An Assessment of the Fosberg and Ellenberg Methods of Classifying Vegetation for Conservation Purposes," *Biological Conservation* 6, no. 1 (1974).

43. E.g., see Peter Galison and Bruce Hevly, eds., *Big Science: The Growth of Large-Scale Research* (Stanford: Stanford University Press, 1992), 291; Jürgen Renn, ed., *The Globalization of Knowledge in History*, vol. 1, Max Planck Research Library for the History of Development of Knowledge Studies (Berlin: Epubli, 2012), 570. For a comprehensive history of ICSU, see Frank Greenaway, *Science International: A History of the International Council of Scientific Unions* (Cambridge: Cambridge University Press, 2006).

44. Edward in Agatha C. Hughes and Thomas P. Hughes, *Systems, Experts, and Computers: The Systems Approach in Management and Engineering, World War II and After* (Boston: MIT Press, 2011), 243.

45. Robert Boote, "Obituary. Max Nicholson: The Prime Mover of the Nature Conservancy and the World Wildlife Fund, Who Helped Inspire Nature Reserves and Ecological Research," *The Guardian*, 28 April 2003, accessed 24 March 2019, http://www.theguardian.com/news/2003/apr/28/guardianobituaries.highereducation; E. Max Nicholson, "Unpublished Autobiography," n.d., ca. 2000–2003, AL EMN: Boxes A.7–A.12.

46. E.g., Eugene P. Odum, "The Strategy of Ecosystem Development," *Science*, n.s., 164, no. 3877 (1969).

47. Chunglin Kwa, "Representations of Nature Mediating between Ecology and Science Policy: The Case of the International Biological Programme," *Social Studies of Science*

17, no. 3 (1987); Golley, *A History of the Ecosystem Concept in Ecology*; Schleper, "Conservation Compromises."

48. Höhler, *Spaceship Earth in the Environmental Age*, 56–57.

49. This led to a special issue of *Scientific American* in 1970; G. Evelyn Hutchinson et al., *The Biosphere*, Scientific American book series (New York: W. H. Freeman, 1970).; also see Joel B. Hagen, *An Entangled Bank: The Origins of Ecosystem Ecology* (New Brunswick: Rutgers University Press, 1992), 64.

50. UNESCO, ed., *Intergovernmental Conference of Experts on the Scientific Basis for Rational Use and Conservation of the Resources of the Biosphere, Paris, 4–13 September 1968: Recommendations* (Paris: UNESCO, 1968); Harold J. Coolidge, "World Biosphere Conference: A Challenge to Mankind," *IUCN Bulletin*, n.s., 2, no. 9 (1968).

51. Gerardo Budowski, "The Biosphere to Come," in *The Environmental Future: Proceedings of the First International Conference on Environmental Future, Held in Finland from 27 June to 3 July 1971*, ed. Nicholas Polunin (London: Palgrave Macmillan 1972), 581ff.

52. Holdgate, *The Green Web*, 109.

53. Budowski, "The Biosphere to Come," 585.

54. Ibid.

55. Ibid., 586.

56. E.g., IUCN, ed., *Proceedings of the Tenth General Assembly, New Delhi 1969* (Morges: IUCN, 1969).

57. In 1971 the UN launched its Second Development Decade. The project, which lasted until 1980, was to mark a new approach to international development that, compared to immediate postwar programs, put more emphasis on the political and economic autonomy of postcolonial and "developing" governments. United Nations General Assembly, "2626(XXV). International Development Strategy for the Second United Nations Development Decade," 1970, accessed 24 March 2019, http://www.un.org/en/ga/search/view_doc.asp?symbol=A/RES/2626(XXV).

58. Lars Emmelin, "The Stockholm Conferences," *Ambio* 1, no. 4 (1972); Wade Rowland, *The Plot to Save the World: The Life and Times of the Stockholm Conference on the Human Environment* (Madison: Clarke, Irwin, 1973).

59. UNESCO, *International Technical Conference on the Protection of Nature: Lake Success, 22–29 August 1949*, 133, 57.

60. E. Max Nicholson, "International Economic Development and the Environment," *Journal of International Affairs* 24, no. 2 (1970).

61. The institute changed its name to the International Institute for Environment and Development (IIED) in 1973.

62. E. Max Nicholson to Duncan Poore, 3 March 1970, LSA EMN/IBP: Box 2, Folder "Relation with IUCN 1963–1972."

63. Raymond F. Dasmann, John P. Milton, and Peter H. Freeman, *Ecological Principles for Economic Development* (New York: John Wiley & Sons, 1973).

64. Raymond F. Dasmann, "Conservation and Rational Uses of the Environment," *Nature and Resources* 4; *African Game Ranching* (Oxford: Pergamon Press, 1964); Dasmann and Jarrell, *Raymond F. Dasmann: A Life in Conservation Biology*, 11.

65. Dasmann, Milton, and Freeman, *Ecological Principles for Economic Development*, 83.

66. John McHale, *The Ecological Context* (New York: G. Braziller, 1970); *The Future of the Future* (New York: G. Braziller, 1969).

67. Peter M. Haas, "Constructing Environmental Conflicts from Resource Scarcity," *Global Environmental Politics* 2, no. 1 (2002); compare Donella H. Meadows et al., *The Limits to Growth: A Report for the Club of Rome's Project on the Predicament of Mankind* (New York: Universe Books, 1972).

68. John McCormick, *Reclaiming Paradise: The Global Environmental Movement* (Bloomington: Indiana University Press, 1991), 78.

69. Stephen Macekura, *Of Limits and Growth: The Rise of Global Sustainable Development in the Twentieth Century* (Cambridge: Cambridge University Press, 2015), 114.

70. E.g., see Nick Cullather, *The Hungry World: America's Cold War Battle against Poverty in Asia* (Cambridge, MA: Harvard University Press, 2011), 252; Mark Gibson, *The Feeding of Nations: Redefining Food Security for the 21st Century* (New York: CRC Press, 2016), 258; James E. McClellan and Harold Dorn, *Science and Technology in World History: An Introduction* (Baltimore: Johns Hopkins University Press, 2008), 420; John C. Avise, *Conceptual Breakthroughs in Evolutionary Genetics: A Brief History of Shifting Paradigms* (London: Academic Press Elsevier, 2014), 67; Jan Sapp, *Genesis: The Evolution of Biology* (Oxford: Oxford University Press, 2003), 199.

71. Robin Pistorius, *Scientists, Plants and Politics: A History of the Plant Genetic Resources Movement* (Rome: International Plant Genetic Resource Institute, 1997), 62ff.

72. Hutchinson et al., *The Biosphere*; Höhler, *Spaceship Earth in the Environmental Age*; Raymond F. Dasmann and UNESCO, *Planet in Peril: Man and the Biosphere Today* (New York: World Pub., 1972); Nicholson, *The Environmental Revolution*.

73. Duncan Poore and IUCN, *Ecological Guidelines for Development in Tropical Rain Forests* (Morges: IUCN, 1976), 8–17.

74. IUCN, WWF, and UNEP, *World Conservation Strategy* (Gland: IUCN, 1980).

75. E.g., Wöbse, *Weltnaturschutz*.

76. E.g., IUCN, *Proceedings of the Tenth General Assembly, New Delhi 1969,* 82.

77. Poul Villaume, Rasmus Mariager, and Helle Porsdam, eds., *The "Long 1970s": Human Rights, East-West Détente and Transnational Relations* (London: Routledge, 2016).

78. E.g., J. Brooks Flippen, "Richard Nixon, Russell Train, and the Birth of Modern American Environmental Diplomacy," *Diplomatic History* 32, no. 4 (2008); Jacob D. Hamblin, "Environmental Diplomacy in the Cold War: The Disposal of Radioactive Waste at Sea During the 1960s," *International History Review* 24, no. 2 (2002).

79. Lino Camprubí, "Review: The Invention of the Global Environment," *Historical Studies in the Natural Sciences* 42, no. 2 (2016); Höhler, *Spaceship Earth in the Environmental Age*; Naomi Oreskes and John Krige, *Science and Technology in the Global Cold War* (Cambridge, MA: MIT Press, 2014); John R. McNeill and Corinna R. Unger, eds., *Environmental Histories of the Cold War* (Cambridge: Cambridge University Press, 2010).

80. E.g., see Katrina Dean et al., "Data in Antarctic Science and Politics," *Social Studies of Science* 38, no. 4 (2008).

81. J. P. Singh, *United Nations Educational, Scientific, and Cultural Organization (UNESCO): Creating Norms for a Complex World* (London: Routledge, 2010); Irena Kozymka, *The Diplomacy of Culture: The Role of UNESCO in Sustaining Cultural Diversity* (New York: Palgrave Macmillan 2014); Frédéric Bosc, *Tell Me About UNESCO* (Paris: UNESCO, 2001), 11; Laura E. Wong, "Relocating East and West: UNESCO's Major Project on the Mutual Appreciation of Eastern and Western Cultural Values," *Journal of World History* 19, no. 3 (2008): 372.

82. Ross B. Talbot, *The Four World Food Agencies in Rome* (Iowa City: Iowa State University Press, 1990), 16; Vladimir Sokolov, "The Biosphere Reserve Concept in the USSR," *Ambio* (1981): 99.

83. Kai F. Hünemörder, *Die Frühgeschichte der globalen Umweltkrise und die Formierung der deutschen Umweltpolitik (1950–1973)* (Stuttgart: Franz Steiner Verlag, 2004).

84. IUCN, *Proceedings of the Ninth General Assembly, Lucerne 1966*, 101.

85. IUPN, *Proceedings and Reports of the Second Session of the General Assembly, Brussels 1950* (Brussels: IUPN, 1951), 47. For discussions on the scientific internationalism of the postwar period, see Paul N. Edwards, "Meteorology as Infrastructural Globalism," in *Global Power Knowledge: Science and Technology in International Affairs*, ed. John Krige and Kai-Henrik Barth (Chicago: Chicago University Press, 2006), 229ff; Rebecka Lettevall, Geert J. Somsen, and Sven Widmalm, eds., *Neutrality in Twentieth-Century Europe: Intersections of Science, Culture, and Politics after the First World War* (London Routledge, 2012), 9.

86. IUCN, *Proceedings of the Ninth General Assembly, Lucerne 1966*, ii, 101.

87. IUCN, *Proceedings of the Twelfth General Assembly, Kinshasa 1975* (Morges: IUCN, 1976), 164.

88. Gerardo Budowski, Frank Nicholls, and Raymond F. Dasmann, "Draft Programme and Budget for 1973–1975," in *Proceedings of the Eleventh General Assembly, Banff, Alberta, Canada 1972*, ed. IUCN (Morges: IUCN, 1972); Duncan Poore, "Progress Report on the Strategy and Its Component Programmes," in *Proceedings of the Thirteenth (Extraordinary) General Assembly, Geneva 1977*, ed. IUCN (Morges: IUCN, 1977).

89. IUCN, *Proceedings of the Seventh General Assembly, Warsaw 1960*, 41.

90. E. Max Nicholson to Gerardo Budowski, 16 January 1969, RSA SCIBP: NHM, Box 1, Folder "First GA SCIBP 1964"; Doubravka Olsakova, "The International Biological Program in Eastern Europe: Science Diplomacy, Comecon and the Beginning of Ecology in Czechoslovakia," *Environment and History* 24, no. 4 (2018). Influential IUCN figures such as Nicholson or Budowski are known to have had some sympathy for Soviet scientific planning and belonged to what has been described as a generation of "technocratic internationalists"; see Johan Schot and Vincent Lagendijk, "Technocratic Internationalism in the Interwar Years: Building Europe on Motorways and Electricity Networks," *Journal of Modern European History* 6, no. 2 (2008); Glenda Sluga, "UNESCO and the (One) World of Julian Huxley," *Journal of World History* 21, no. 3 (2010); Gerardo Budowski, "Should Ecology Conform to Politics?," *IUCN Bulletin*, n.s., 5, no. 12 (1974): 46; Nicholson, *The Environmental Revolution*.

91. Anna Medwecka-Kornas to E. Max Nicholson, 16 May 1967, RGSA EMN: Box 4; Max Nicholson to Gerardo Budowski, 16 January 1969, RSA SCIBP: NHM, Box 1, Folder "First GA SCIBP 1964."

92. De Bont, Schleper, and Schouwenburg, "Conservation Conferences and Expert Networks in the Short 20th Century."

93. Due to a change in recording, counts for the General Assemblies of 1960, 1963, and 1966 took place at the beginning of the gathering. The later counts took place toward the end of the meetings and already included those members newly elected at the occasion of the Assembly. The data is taken from the proceedings of the regular IUCN General Assembly meetings between 1960 and 1981. IUCN, *Proceedings of the Seventh General Assembly, Warsaw 1960*; *Proceedings of the Eighth General Assembly, Kenya*

1963; *Proceedings of the Ninth General Assembly, Lucerne 1966*; *Proceedings of the Tenth General Assembly, New Delhi 1969*; *Proceedings of the Eleventh General Assembly, Banff, Alberta, Canada 1972*; *Proceedings of the Twelfth General Assembly, Kinshasa 1975*; *Proceedings of the Fourteenth General Assembly, Ashkhabad 1978* (Morges: IUCN, 1979); *Proceedings of the Fifteenth General Assembly of IUCN, Christchurch 1981* (Gland: IUCN, 1983). The Thirteenth Extraordinary General Assembly of 1977 did not entail an election and was therefore not included.

94. Gerardo Budowski and Frank Nicholls, "Executive Board: Progress of IUCN Expansion Plans," HUA HJC: HUG (FP), 78.20 Box 1, Folder "IUCN-WWF and Related International Conservation 1970 IUCN-HJC President-Executive Board-May meeting."

95. C. M. Berkeley to Tracy Philips, 16 March 1956, UNESCO IUCN: Box 502.7, Folder A 01 IUCNNR-6.

96. Udall cited in IUCN, *Proceedings of the Eighth General Assembly, Kenya 1963*, 47.

97. De Bont, Schleper, and Schouwenburg, "Conservation Conferences and Expert Networks in the Short 20th Century."

98. Holdgate, *The Green Web*, 115.

99. Boardman, *International Organization and the Conservation of Nature*, 98.

100. Kozymka, *The Diplomacy of Culture*; Marc Frey, Sönke Kunkel, and Corinna R. Unger, eds., *International Organizations and Development, 1945–1990* (London: Palgrave Macmillan, 2014); Ahmend Mahiou, "Declaration on the Establishment of a New International Economic Order," UN Audiovisual Library of International Law, 2011, accessed 23 March 2019, http://legal.un.org/avl/pdf/ha/ga_3201/ga_3201_e.pdf.

101. E.g., Giuliano Garavini and Richard R. Nybakken, *After Empires: European Integration, Decolonization, and the Challenge from the Global South 1957–1986* (Oxford: Oxford University Press, 2012), 30ff.

102. UN, *Development and Environment (Subject Area V)*, United Nations General Assembly (New York: UN, 1971).

103. Louis Emmerij, Richard Jolly, and Thomas G. Weiss, *Ahead of the Curve? UN Ideas and Global Challenges* (Bloomington: Indiana University Press, 2001), 60.

104. Commission on International Development and Lester B. Pearson, *Partners in Development: Report* (New York: Praeger, 1969); Matthias Schmelzer, "The Club of Rome to Help the Poor? The OECD, "Development," and the Hegemony of Donor Countries," in *International Organizations and Development, 1945–1990*, ed. Marc Frey, Sönke Kunkel, and Corinna R. Unger (London: Palgrave Macmillan, 2014), 195.

105. Vijay Prashad, *The Darker Nations: A People's History of the Third World* (New York: New Press, 2008), 132ff.; Sönke Kunkel, "Contesting Globalization: The United Nations Conference on Trade and Development and the Transnationalization of Sovereignty," in *International Organizations and Development, 1945–1990*, ed. Frey, Kunkel, and Unger, 247–48; also see Macekura, *Of Limits and Growth*, 223.

106. E.g., Adam, *Elephant Treaties*; Tilley, *Africa as a Living Laboratory*. "Charismatic" is a common term in conservation biology, used to describe animals with symbolic value and popular appeal.

107. IUCN, "General Statement (Revised): IUCN's African Special Project (ASP) 1960–1963," LSA EMN/IBP: Box 8, Folder "CCTA/IUCN Arusha 5–12 September 1961"; IUCN, FAO, and UNESCO, *CCTA/IUCN Symposium on the Conservation of Nature and Natural Resources in Modern African States (in Collaboration with the FAO and*

UNESCO), Arusha, Tanganyika, 5–12 September 1961 (Lagos: Commission for Technical Cooperation in Africa South of the Sahara, 1962); IUCN, *Proceedings of the Eighth General Assembly, Kenya 1963*, 63; Macekura, *Of Limits and Growth*, 47; Alexander B. Adams, IUCN, and UNESCO, eds., *First World Conference on National Parks: Proceedings of a Conference, Seattle, Washington, June 30–July 7, 1962* (Washington, DC: National Park Service, U.S. Department of the Interior, 1964); Interview with Lee Talbot, 11 September 2014.

108. Talbot and IUCN, *A Look at Threatened Species.*

109. IUCN, ed., *Proceedings of the International Meeting on the Use of Ecological Guidelines for Development in the American Humid Tropics, Held at Caracas, Venezuela, 20–22 February 1974* (Morges: IUCN, 1975); *Proceedings of a Regional Meeting on the Use of Ecological Guidelines for Development in the Tropical Forest Areas of South East Asia, Held at Bandung, Indonesia 29 May to 1 June 1974* (Morges: IUCN, 1975).

110. IUCN, ed., *Proceedings of the Thirteenth (Extraordinary) General Assembly, Geneva 1977* (Morges: IUCN, 1977), 15.

111. Holdgate, *The Green Web*, 198.

112. IUCN, *Proceedings of the Twelfth General Assembly, Kinshasa 1975*, 180.

113. Ibid., 50.

114. IUCN, *IUCN Yearbook: Annual Report* (Morges: IUCN, 1975).

115. Rowland, *The Plot to Save the World: The Life and Times of the Stockholm Conference on the Human Environment*, 35–37.

116. IUCN, *Proceedings of the Fourteenth General Assembly, Ashkhabad 1978.*

117. IUCN, *Proceedings of the Thirteenth (Extraordinary) General Assembly, Geneva 1977*, 51A–B.

118. Mohamed Kassas, "Address at the Opening of the 15th Session of the IUCN General Assembly," in *Proceedings of the Fifteenth General Assembly of IUCN, Christchurch 1981*, ed. IUCN, 86.

119. For a discussion of American neo-Malthusianism during the 1970s, see Adams, *Against Extinction*, 51ff., 129ff.; *Green Development*, 142.

120. Macekura, *Of Limits and Growth*, 35.

121. John Krige, *American Hegemony and the Postwar Reconstruction of Science in Europe* (Cambridge, MA: MIT Press, 2006).

122. IUCN, *Proceedings of the Fourteenth General Assembly, Ashkhabad 1978.*

123. Holdgate, *The Green Web*, 252.

124. Ossamma M. El-Tayeb, "Internal Memorandum," to Reuben Olembo, UNEP Senior Programme Officer, 30 March 1979, cited in Holdgate, *The Green Web*, 151.

125. Jon Tinker, "World Environment: What's Happening at UNEP?," *New Scientist* 66, no. 953 (1975); Tony Loftas, "The UN's Agents of Change," *New Scientist* 66, no. 953 (1975); Stanley Johnson, *UNEP: The First 40 Years; A Narrative* (Nairobi: UNON/Publishing Section Service, 2012), 14ff.

CHAPTER 2

Classifying Ecosystems
The International Biological Program, 1964–1974

Introduction

From 1964 until 1974, the International Biological Program (IBP) brought together ecologists, botanists, and zoologists from around the world to study both the productivity and vulnerability of natural ecosystems.[1] Ideas for such an international science program on biology had emerged within the framework of the International Union of Biological Sciences (IUBS) around 1960. From the very beginning, these developments were followed with much interest by many conservationists at the International Union for Conservation of Nature and Natural Resources (IUCN). Never before had there been an international program in the life sciences, and never before had ecology received so much attention. IUCN's president, Jean Georges Baer, a prominent zoologist and a member of the IUBS himself, sensed the opportunity to enhance IUCN's scientific and financial status.[2] Although the original plans and their initiators had relatively little to do with traditional nature protection, IUCN conservation experts gained a foothold in the IBP's preparation process relatively early on.[3]

In May 1962, Baer successfully maneuvered an important IBP planning meeting to the IUCN headquarters in Morges, Switzerland. The meeting proved decisive for putting nature conservation on the IBP's agenda. When in 1963 the IBP was formally launched in Vienna at the Tenth General Assembly of the International Council of Scientific Unions (ICSU), an IBP Section for the Conservation of Terrestrial Communities (IBP/CT) was established as one of seven international subcommittees.[4] Additionally, the distribution of executive positions for the IBP turned out advantageous for scientists interested in the conservation of nature. At the first official meeting in Paris in 1964, Baer was elected president of the IBP's scientific board, the Special Committee of the IBP (SCIBP). After further discussions, the scientific directorship of SCIBP was offered to Barton Worthington from the British Nature Conservancy, one of IUCN's long-standing member organizations (figure 2.1). Another British conservationist, Max Nicholson, was appointed convener of the IBP/CT section. With several leading IUCN conservationists in high-ranking

Figure 2.1. The last meeting of the SCIBP at the Royal Society, London, 1974. Reproduced from LSA EMN/IBP: Box "Various Mags with Articles on IBP (1960–1970)."

Left to right: in the back, François Bourlière, Giuseppe Montalenti; *second row,* Livia Tonolli, Gina Douglas, Joseph Weiner, Ronald Keay; *below them,* Max Dunbar, Susan Darrell-Brown; *third row,* Frank Blair, Makoto Numata, Mike Baker of ICSU, Malcolm Alan Winter of Cambridge University Press; *front row,* Barton Worthington, George Davis, Max Nicholson, Jean G. Baer, Cyril Waddington, Michel Batisse (UNESCO), Sir Otto Frankel, either B. R. Sechachar or Gopal Ayengar, Jan Květ, Terrence Moore of Cambridge University Press, Sir Hugh Elliott of IUCN. Identified with the help of Gina Douglas.

positions in the IBP, the future looked promising for a strong conservation element in the program.

The focus of this chapter is on the role of the IBP in science-based and internationally organized nature conservation. The IBP had been inspired by advancements in the geosciences brought about by the International Geophysical Year of 1957. Historically, it fell into a period of large-scale science programs and of planning and reconstruction schemes in the postwar era.[5] The IBP had been founded against the background of growing concerns among natural scientists about the finiteness of the planet's natural resources and growing population pressures. This chapter, then, analyzes how IUCN conservationists linked their ideas on the protection of nature to the IBP, and, in turn, what their venture into big biology meant for nature conservation as a scientific and internationally recognized field. In the few but valuable accounts on the history of the IBP, historians of science such as Chunglin Kwa, Frank Golley, and more recently David Coleman have pointed out that the IBP constituted a formative point in the history of ecology as a discipline.[6] This chapter shows how the same is true for the course of internationally organized nature conservation. The IBP changed conservationists' approach to ecology as well as their ideas on the global implementation of conservation measures.

During the IBP, a rift occurred within the international machinery of nature protection. At the heart of the dispute was the type of ecology to constitute the scientific underpinning of international conservation work and the ways in which conservation measures were to be implemented around the globe. One group of IUCN experts coalesced around the IBP/CT convener Nicholson. They pushed for the scientific recognition of their expert roles by linking conservation to the emerging field of cybernetically inspired ecosystems ecology. Promoters of this type of ecology looked at the natural processes in ecosystems and aimed at calculating the potential productivity of these ecological systems with the help of computers. At about the same time, a second group began to form at the executive center of IUCN, around Director General Gerardo Budowski and Senior Ecologist Ray Dasmann. In contrast to Nicholson and his colleagues, this second group was not interested in cybernetic calculations on processes or productivity. Instead, they based their notion of conservation on descriptive community ecology for the purpose of mapping threatened species in particular regions. Whereas Nicholson favored a centralized approach, independent from regional politics, Dasmann promoted a more decentralized line—aiming at a stronger liaison with the special agencies of the United Nations (UN) to benefit from their contact with national governments and existing local research projects.

By the mid-1970s, the discussions between both groups had reformed the methods, objectives, and self-presentation of international science-based conservation in significant ways. In fact, out of the dispute, a new shared conservation approach emerged, which focused on the protection and study of ecosystems in different biogeographical regions. With this new approach, involved conservationists were able to balance a set of partly antithetic requirements for the design and implementation of international environmental projects. On the one hand, it allowed conservationists to promote the universal significance and global applicability of ecological guidelines for all questions related to the living natural environment. On the other hand, it granted the flexibility demanded by UN agencies for local decision-making and action priorities when it came to the protection of the plant and animal resources in particular countries.

Ecosystem Ecology and the International Organization of Conservation

The IBP was generally perceived as an opportunity for the conservation community to prove their scientific expertise and the international relevance of their work. Furthermore, in the 1960s, many IUCN conservationists regarded their participation in the IBP as a potential instrument to further expand the scope of their work and to advance the recognition of the Union as a scientific organization. There were, however, different ideas about how the IBP would contribute to international and science-based conservation and about what this would mean for the work of established organizations such as IUCN. Early ideas on a big science program for biology evoked diverse reform plans among IUCN conservationists regarding the scientific foundation and the international organization of their work.

A first point of discussion was the IBP's focus on ecosystem ecology. By 1960, many conservationists within IUCN were working with the concept, but not, however, in a unified way. Also in their discussion on what the IBP could mean for conservation science, conservationists differed in their interpretation of ecosystem ecology and how it should be applied in conservation advice. In one interpretation, the IBP's focus on ecosystems was seen as an opportunity to strengthen the work on the protection of threatened species and populations as already pursued within IUCN. As early as 1960, Baer had suggested at a planning meeting in Lisbon that the study of immediately threatened biological communities and the ecosystems they resided in presented a suitable topic for an International Biological Program.[7] The science-based protection of species and communities was firmly rooted within IUCN's program. The organization had started with the collection of data on threatened species as early as the 1940s. By the mid-1960s, the *Red List of Threatened Species* by the Species Survival Commission had become a cornerstone in the Union's program.[8] Lists on threatened species and their protection dated back to nineteenth-century biogeographical practices of mapping and recording species present in particular landscapes or areas. Since the mid-twentieth century, population and community ecologists had linked this kind of data to specific soils and climates. The focus of these studies in particular was on the development of pristine communities of different species populations, believed to reach a naturally balanced state if left alone in their specific environments.[9] Baer's suggestion to make threatened ecosystems the core of IUCN's involvement in the IBP did not, however, satisfy everyone within the organization.

Since the early 1950s, a number of IUCN members had been demanding a broader and more universal scope of, and more general guidelines for, conservation work. That more inclusive definitions of conservation, concerned with environmental problems at large, began to overshadow narrow preservationist ones was indicated in 1956 by the name change to the International Union for Conservation of Nature and Natural Resources.[10] For some conservationists, the IBP's proclaimed focus on ecosystem ecology promised an opportunity to expand and standardize the work of the Union in accordance with its new name.[11] For this they drew on the work of the American ecologists and brothers Eugene and Howard Odum. In the mid-1950s, the Odum brothers had developed a new holistic approach to ecosystem ecology that was inspired by early computer technology and cybernetics. Its particular focus was the cycles of productivity or flow of nutrients within such systems and their development over time.[12] These ideas were picked up by the IBP's initiators as a response to the rise of more reductionist molecular sciences in the 1950s and early 1960s, which threatened to outcompete botany, zoology, and ecology projects in status and financial support.[13] It was hoped the program of the IBP could contribute to the formulation of predictive ecological theories that could replace the rather descriptive community ecology of the recent past.[14]

Next to the scientific basis, the international scope of the IBP was perceived as a chance to strengthen IUCN's international support. When ideas for an IBP emerged, its international relevance was never questioned. Like the members of many other large science programs or organizations, IUCN conservationists perceived international cooperation to be beneficial in terms of prestige and funding.[15] Yet, different ideas coexisted with regard to how conservation work should be managed on a global scale. There were different ways of organizing such an international cooperation, based on different conceptualizations of scientific internationalism. One option was to cooperate more strongly with the work of the UN's intergovernmental agencies. In recent years, UNESCO, one of IUCN's first sponsors, had developed a strong and financially secure research program on natural resources. Already in 1960, UNESCO had reorganized their natural science projects in a new Division of Studies and Research relating to Natural Resources.[16] Within this division, nature conservation was featured alongside studies on hydrology, geology, soil science, and other ecological studies. The program, which started its operational phase parallel to the IBP launch in 1964, had a budget of nearly US$800,000 for the first two years alone. This was a substantial sum compared to the modest US$200,000 awarded on average each year for all IBP projects combined.[17] UNESCO got involved in the SCIBP as a sponsor and advisor, organizing joint meetings and financing about 15 percent of the IBP's total budget.[18]

At the time, IUCN conservationists were already entertaining links with scientists at UNESCO.[19] In fact, UNESCO's Advisory Committee consisted of fifteen internationally known specialists, many of whom were also members of IUCN. For a number of leading IUCN conservationists, the future seemed to promise a closer cooperation between the two organizations. This not only held potential financial benefits for IUCN, but also the chance to expand the programmatic similarities between the two organizations. A large fraction of the Union's efforts at the time focused on traditional, formerly colonial conservation areas in Africa and Asia.[20] Similarly, the conservation issues that UNESCO had so far been concerned with included programs on arid lands and the humid tropics. These aimed at giving advice for particular localities.[21] The work of UNESCO, based on a representation of member states, seemed to promise conservationists additional channels to reach government officials and poli-cymakers otherwise not included in IUCN. At the beginning of the IBP, then, there was a fraction of conservationists looking for ways to strengthen IUCN's already existing links with UNESCO even further.[22]

Yet for IUCN, working with the more affluent intergovernmental organi-zations would also mean playing by their rules, something that concerned a number of conservationists in IUCN. Agencies such as UNESCO were at times affected by the interests of national politics, as they depended on the cooperation of its member states. This could in turn inflict problems for in-ternational scientific projects. Such problems occurred in 1957, when, much to the dismay of many conservationists, South Africa, with its vast plains and migrating herds, and one of the regions linked to early international nature conservation efforts, left UNESCO because its government felt UNESCO's impartiality statement interfered with the domestic politics of the Apartheid regime.[23] Weary of these kinds of political problems already in the late 1940s, IUCN's founding members had emphasized their independence from UN membership politics.[24] By 1960, however, IUCN itself had developed into a complicated nexus of diverse national agencies and NGOs with different opin-ions on what kind of nature needed protection and how conservation should be implemented. When plans for the IBP became known, a number of IUCN members hoped that new links with the nongovernmental ICSU would mean new political independence for IUCN and a new focus on science rather than bureaucracy.[25] For those scientists who were discontent that the focus of many IUCN and UNESCO projects was on the Southern Hemisphere alone, a big biology program under the auspices of ICSU also promised the expansion of conservation in geographical scope.

Many members of ICSU had understood science all along as something in-herently global and apolitical. The organization propagated a universalist leit-motif, according to which "academies might sit down side by side even when governments could not."[26] In the postwar period, left-leaning internationalists, such as the British ecologist Julian Huxley, continued to promote this form of thinking. In the 1950s, this universalist ideology, working on one centrally decided set of problems, also manifested itself in the International Geophys-ical Year, ICSU's first international science program that was global in scope, which had been inspired by the launch of the first space satellite in 1957.[27] The program employed scientists from different world regions, irrespective of ex-isting political tensions, who were working through national academies of sci-ence rather than governments. Unlike UNESCO, for instance, in the 1960s and 1970s, ICSU would accommodate scientists from both Germanys and both Chinas, as well as scholars from Israel. Western ICSU scientists kept good rela-tions with colleagues from Russia and the USSR throughout the Cold War, and South Africa continued its ICSU membership when the country left UNESCO after controversies concerning the apartheid regime. A similar aura of universalist science and a unified scientific community surrounded the plans for the IBP when in the early 1960s the ICSU managed to avoid a visa ban for non-members of the North Atlantic Treaty Organization (NATO). This triumph over early Cold War politics seemed to confirm that scientific NGOs such as ICSU presented a viable alternative to the UN's intergovernmental model for research cooperation, by then caught up in diverse political agendas.[28] For some conservationists at IUCN then, the IBP also presented a tool to promote the universal rule of ecology-based nature protection, irrespective of national ideologies or political agendas.

By the mid-1960s, it was apparent that the conservationists directly involved in the IBP/CT section intended to reform nature conservation according to a broader and more universalist interpretation. For this they strongly banked on their status as independent scientists—a status granted to them by their membership in the IBP and the ICSU. On 14 October 1965, the members of the section, staffed with an international IUCN elite, published their program in the *New Scientist*.[29] The IBP/CT members not only announced their plans in a leading science journal, but also fashioned them in the style of a global research project. Ecological investigation in all world regions was presented as both the means and end to science-based nature protection.[30] This endeavor constituted a clear move away from conservation concerned with saving threatened ecosystems in particular regions. A "world network of research reserves" was to provide the research areas necessary for the relevant scientific studies.[31] In 1965, American ecologist Edward Graham, deputy convener of the IBP/CT section and chairman of IUCN's Commission on Ecology, pronounced in the *New Scientist*:

> The study of natural ecosystems holds the answer to many outstanding questions in biology, both pure and applied. Such complexes, however, are being destroyed at unprecedented rate. The Section on Conservation of Terrestrial Communities of the International Biological Programme is to work at the establishment of worldwide reserves, representing characteristic ecosystems.[32]

For this long-term goal, IBP/CT conservationists first needed to know more about the current condition and protection status of the world's ecosystems. For this purpose, the IBP/CT section designed a classification and check-sheet survey. The project offered an opportunity to those conservationists within the IBP/CT and IUCN who wanted to carry out more fundamental and quantitative research into the scientific conservation of ecosystems.[33]

At the same time, conservationists at the IBP suggested that such a wide-ranging ecological survey, unprecedented at the time, would complement and add to the work of IUCN without creating programmatic conflicts between the two organizations.[34] Very soon, however, frictions emerged between IUCN conservationists in the IBP demanding more fundamental ecosystem research, and those IUCN members who insisted that ecosystem conservation was about saving threatened species in their particular, local habitats. These frictions became more tangible when two alternative programs for ecosystem conservation emerged within IUCN. On the one hand, IUCN conservationists working for the IBP/CT began to develop a classification and a check-sheet survey to identify and classify different ecosystem types around the world—basic research that they hoped to inject into IUCN's projects at a later stage.[35] On the other hand, a number of fellow IUCN conservationists suggested a closer cooperation with UNESCO. Here, attempts to design a system for mapping the world's vegetation types were already underway, and these, so it was hoped, would help to identify and locate endangered ecosystems.

At first sight, the differences between the two classification systems might look insignificant. Both mainly focused on vegetation as an indicator for ecosystem types, and both were to be used in different world regions. In fact, however, the two classification systems epitomized two distinct approaches to international conservation. An examination of these approaches in more detail reveals that they were not based on discrepant ecological theories alone. They also demonstrated crucial differences regarding the kind of nature they aimed to protect, the role assigned to conservation experts, and the ways in which international nature protection was to be implemented.

Predicting Productivity with the IBP/CT Check Sheets

In order to understand the rationale behind the classification and check-sheet survey designed by the members of the IBP/CT section, it makes sense to look at their main promoter and the section's convener, Max Nicholson. Nicholson, who regarded the CT section as his own brainchild, enjoyed the privilege of appointing his own section members and set in motion much of its program. Nicholson, born in 1904 in Ireland to English parents, was not a biologist by training, but came to ecology through his interest in bird-watching.[36] In the late 1920s, this interest was fostered by his extracurricular activities at the Oxford Exploration Club, where Nicholson met several influential ecologists, including Julian Huxley, Arthur Tansley, and Charles Elton. Elton's population biology had a particularly strong influence on Nicholson's ecological worldview. In this worldview, Nicholson conceptualized conservation as nature management. Elton's theory of population dynamics opposed the predominant idea of a given balance in nature.[37] In line with these ideas, Nicholson also did not believe in a superior, natural stability in the absence of human influences. In contrast, Nicholson thought the conservationist's task was to

create and manage harmonious, natural and nonnatural systems. His understanding that managing nature by manipulating external environmental factors was not only possible but necessary to guarantee a balanced system was
crucial to Nicholson's ecological and conservationist writings and his plans for
the IBP/CT.

While Nicholson learned his ecology from Elton, the more social science–
oriented ecologists, like Tansley and Huxley, had a significant impact on his
view on human society and humankind's place in nature. In contrast to Elton, who had confined his work to environmental influences on animal populations, Tansley and Huxley applied ecology to wider societal problems.[38]
Crucial in the early work of Tansley and Huxley was a temporal, evolutionary
dimension that included a diachronic account of humanity's relation to the
living environment. At the same time, their idea of ecology also entailed a view
of the future in which every generation was responsible for maintaining the
balance between nature and society for the generations to come. Huxley, who
in the late 1920s had visited the USSR and admired the results of large-scale
social and economic planning, especially influenced Nicholson's ecological
worldview and his technocratic tendencies.[39]

These tendencies took a more definite shape in his work at the British Nature
Conservancy. During the 1950s, Nicholson assigned a leading role to scientific
experts and stressed the need for centralized scientific steering committees
when it came to studying and managing different landforms.[40] In line with
Tansley, he saw the border between the natural and the artificial as arbitrary.[41]
In theory, he distinguished between "unconverted" nature—shaped through
natural selection—semimodified areas, and natural ecosystems completely
modified by man.[42] He stressed, however, that untouched nature virtually no
longer existed—neither in Great Britain nor in most other parts of the world.
Accordingly, he thought that conservation approaches that pursued the mere
isolation of ecosystems, without management or research, would contribute
little to environmental problem-solving.[43] For Nicholson, the objective of conservation was essentially to distribute and manage natural resources, wild as
well as farmed.[44]

Although he had successfully implemented this approach at the Nature
Conservancy, Nicholson believed his increasing environmental concerns
called for global solutions. Early on, he had been involved in the activities of
IUPN, and later IUCN. During the 1950s, he was a regular participant in the
meetings of the Commission on Ecology and presented on his experiences
relating to the exchange between ecological research and land management.[45]
However, as a technocrat and centralist planner, Nicholson not only was dis-
contented with the regionally limited approach to conservation of IUCN's
Special Projects in Africa and Asia, but also thought the bureaucratic management of IUCN impeded the efficient implementation of conservation ex-

pertise and demonstrated the lack of clear scientific objectives.[46] In the IBP's overall focus on system ecology, Nicholson saw the chance to unify once and for all the diverse theories behind conservation by devising truly global, coherent, and future-oriented principles for ecosystem conservation and natural resource management.[47]

In particular, Nicholson was intrigued by computational models of energy cycles of biological systems.[48] In the 1950s, the introduction of computers had allowed the study and simulation of functions of any system; at the same time, ecosystem ecology experienced a new upswing.[49] With the development of cybernetic approaches in ecology, Nicholson's demands for universal principles for conservation seemed to find a scientific basis.[50] Computer technology was key to the new method Nicholson envisioned for a type of conservation that went far beyond the preservation of particular species or habitats.[51] The functioning of natural systems, he believed, could not be studied in fragments, but only by looking at the system as a whole. Systems thinking, then, constituted the necessary move away from simple, or locally grounded, cause-and-effect thinking and short-term planning. Problems could not be assessed or tackled locally without taking into account the functioning of the larger system. Like, for instance, the Odum brothers, who created system models for other IBP subprojects, Nicholson commissioned a model on the interdependencies between the "biosphere," the environmental envelope around the planet that made life possible, and the "technosphere," the socioeconomic and industrial system that in his eyes characterized modern human living. The model would lay out the continuous and reciprocal interaction of human society with its environment (figure 2.2).[52] Nicholson thought this system approach offered a holistic outlook toward environmental problems that could back up his technocratic planning approach, linking society, land use, and natural resource management.

With the image of the biosphere and the technosphere, Nicholson illustrated the societal relevance of the work of his IBP/CT section to audiences beyond the biological community. Midway through the IBP/CT's so-called action phase, which started in 1968 and lasted until 1972, he used the model to address development experts at the Columbia University Conference on International Economic Development, organized by the British economist Barbara Ward. The model allowed Nicholson to explain "nature's and man's interaction as value-free physical processes."[53] Designed to function as an objective tool of demonstration, the model nevertheless depicted Nicholson's ideology-laden message that understood society and nature as part of an interdependent, yet manageable system. According to Nicholson, ecological management was needed to avoid undesirable or even disastrous side effects of the human use of nature, such as the overexploitation or the industrial poisoning of ecological systems. Nicholson's wider philosophy remains somewhat hidden in the

Figure 2.2. Diagram of the biosphere and the technosphere, Royal Society Symposium, 1969. Reproduced from LSA EMN/IBP: Box 4, Folder "World Bank etc.," with kind permission of Land Use Consultants, London.

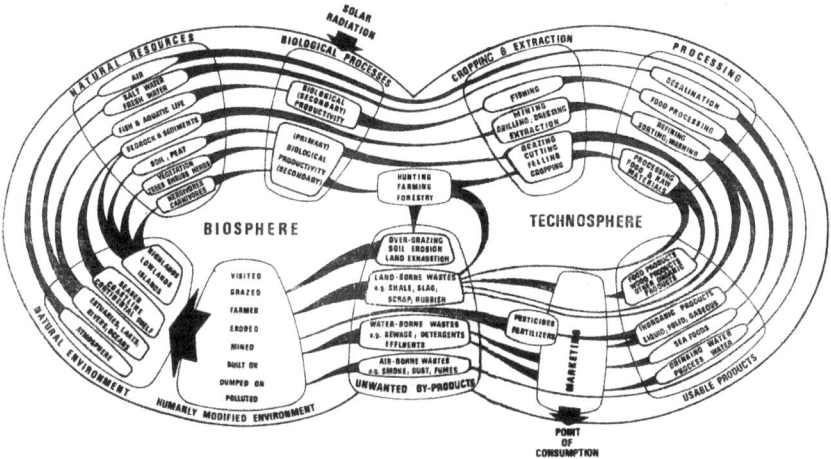

official IBP documents, but was made more explicit in writings he published in the same period. These contain numerous passages where Nicholson expressed his worries about the disturbed and mostly ignored relationship between nature and man, and where he assigned an important role to a holistic systems-based expertise:

> It will not be possible to harmonise human development with the natural environment on the necessary grand scale until those in charge . . . are educated afresh so that they learn to see problems as a whole. . . . Technologists of broader and deeper formation, with complex and well-balanced professional training, will alone be capable of successfully handling the immense problems of adaption now facing us.[54]

This passage clearly illustrates that Nicholson perceived both nature and society as interrelated entities that needed to be balanced and managed by experts.

These key ideas—a system of thinking that encompassed man within nature, future-oriented and global plans, a technocratic role for conservationists, and the top-down implementation of their advice—already structured Nicholson's activities as head of the Nature Conservancy in Britain. In 1963, the British Nature Conservancy, under Nicholson's presidency, carried out a survey on the human impact on the British environment.[55] This led to a chart that indicated different land types affected in time and space, and that listed land development problems as well as possible solutions. In his role as convener of the IBP/CT section, Nicholson planned comparable surveys to be carried out for all world regions. These surveys could help to forecast problems and to provide guidelines of action and protection, especially as he feared that, in the future, a growing world population would only intensify exploitative land-use practices.[56] In Britain, the Nature Conservancy's survey was used to inform economic decision-making on a national level. Nicholson envisioned a similar top-down approach for international development projects in which conservation experts were to take the lead. This same line of thought was behind the IBP's classification and check-sheet survey of the world's ecosystems, which Nicholson initiated. The design of a general ecosystem classification, coherent surveying methods, and the computerization of globally collected ecological data would bring conservation a significant step closer to becoming the superordinate science for global land management that Nicholson intended it to be.[57]

Designing such a classification system, however, required epistemological choices, since a recognized way of classifying ecosystems had yet to be found. In a seamless cybernetic philosophy, it remained undetermined where one ecosystem ended and another started, and this also varied with the type and topography of the concerned area.[58] Nicholson therefore believed some pragmatic initial system had to be devised to reflect "on a universal scale but on a selective manageable basis" which information was necessary to understand the natural environment.[59] In designing such an initial system, Nicholson took his inspiration from Elton, according to whom all energy production cycles within an ecosystem were based on the generation of plant material from solar energy. According to this line of reasoning, vegetation could serve as a reference for ecosystem structures.[60] After several rounds of discussion, the members of the IBP/CT section decided to use the "Classification of Vegetation for General Purposes" by the Smithsonian botanist Raymond Fosberg for the categories of the ecosystem survey they devised (figure 2.3).[61]

In essence, Fosberg's classification featured two decisive characteristics. First, it did not depend on floristic criteria, meaning in this case that it did not record the names of particular plant species. Instead, it aimed to capture the physiognomic structure of the concerned ecosystem. This meant that the classification described vegetational arrangements in space—whether the vegetation was open, as in a forest where the tree canopies did not form a continuous cover; closed, like a field of tall grass covering a savanna; or sparse, like dispersed trees in a steppe. Moreover, the Fosberg classification included information on more general biological processes—for example, whether the trees in the formation were shedding their leaves with the changing seasons.[62] Second, Fosberg purposefully avoided the incorporation of place-specific environmental information such as climatic or geographic details in his classification. Such information, as well as information on the actual plant species

Figure 2.3. Fosberg's classification for the IBP/CT check-sheet survey. Reproduced from Arthur R. Clapham, ed., *The IBP Survey of Conservation Sites: An Experimental Study* (Cambridge: Cambridge University Press, 1980), 251.

Legend:

- ■ Closed
- O Open
- S Sparse
- x Absent closed
- oo Absent open
- s Absent sparse
- (blank) Absent

			Floating aquatic	Submerged aquatic	Bryoid	Broad leaved herbs	Short grass	Tall grass	Dwarf shrub	Shrub	Tree
Closed vegetation											
I	A	Forest			x	x	x	x	x	x	■
	B	Scrub			x	x	x	x	x	■	
	C	Dwarf scrub			x	x	x		■		
	D	Open forest with closed lower layers			├───── Closed ─────┤						O
	E	Closed scrub with scattered trees			x	x	x	x	x	■	S
	F	Dwarf scrub with scattered trees			x	x	x		■		S
	G	Open scrub with closed ground cover			├──── Closed ────┤					O	
	H	Open dwarf scrub with closed ground cover			├── Closed ──┤				O		
	I	Tall savanna			├──── Closed ────┤			s	s	s	S
	J	Low savanna			├── Closed ──┤		s	s	s	s	S
	K	Shrub savanna			├──── Closed ────┤				Sparse		
	L	Tall grass			x	x	x	■			
	M	Short grass			x	x	■				
	N	Broad leaved herb vegetation			x	■	oo	oo			
	O	Closed bryoid vegetation			■	s	s				
	P	Submerged meadows	oo	■							
	Q	Floating meadows	x	■							
Open vegetation											
2	A	Steppe forest			oo	oo	oo	oo	oo	oo	O
	B	Steppe scrub			oo	oo	oo	oo	oo	O	
	C	Dwarf steppe scrub			oo	oo	oo		O		
	D	Steppe savanna			oo	oo	├─ O ─┤		oo	oo	S
	E	Shrub steppe savanna			oo	oo	oo	oo	s	S	
	F	Dwarf shrub steppe savanna			oo	oo	O		S		
	G	Steppe			├────── O ──────┤						
	H	Bryoid steppe			O						
	I	Open submerged meadows	s	O							
	J	Open floating meadows	O	s							
Sparse vegetation											
3	A	Desert forest			s	s	s	s	s	s	S
	B	Desert scrub			s	s	s	s	├─ S ─┤		
	C	Desert herb vegetation			├────── S ──────┤						
	D	Sparse submerged meadows		S							

present, had to be recorded separately, but not as part of the general categories. In this way, Fosberg's system could help to generate fundamental knowledge about different types of ecosystems, without drawing a priori conclusions on links to particular species, regions, or climates.

Nicholson believed this approach to be advantageous for two reasons. On the one hand, as it worked independently of local species occurrences, he believed the Fosberg system could be employed to study ecosystems anywhere on the globe. On the other hand, the check-sheet survey based on Fosberg's classification could be used to do ecological groundwork beyond conservation in the strictest sense. Nicholson hoped that a separate recording of these general ecosystem types and environmental data would eventually allow for objective research on the relations between climates, vegetation structures, and the species that could grow in these ecosystems. In discussion with his IBP staff, Nicholson explained the importance of collecting data in "pure" form, so that "any subsequent attempts to correlate e.g. vegetation occurrence with soil or climate will be entirely free of the suspicion that the method of collecting data contains a built-in bias towards assuming some type of connection [between, for instance, climate, region, and vegetation] which should be the function of the data to test objectively."[63]

The separated data units were recorded on a check sheet, paired with a worldwide grid square reference system, and stored in a punch-card database at Monks Wood in Cambridgeshire, England, a reserve and research station run by the British Nature Conservancy.[64] Nicholson hoped these data units could eventually instruct the anticipatory scientific planner. In line with Elton's principles of population dynamics, Nicholson believed that the organization and structure of biotic communities in different habitats was in fact very similar.[65] Therefore, the data of the check-sheet survey would not only make basic information available to study the functioning of ecosystems and to determine different types of natural systems, but would also enable future computer calculations on the potential productivity of different plant and land type combinations. This way, the data from the check-sheet survey would enable future scientists to forecast which crops, and later, which animals, could potentially prosper in a particular location.[66] Throughout the years of the IBP and its synthesis until 1974, Nicholson was optimistic that this kind of ecological study would reform global conservation and land use practices, coordinated by his team at Monks Wood.[67]

Recording Threatened Ecosystems with the UNESCO Maps

A few years after Nicholson had begun to pursue his ambitious reform plans for international conservation standards, a second group of IUCN conservationists gained prominence. This group recognized in IBP's Conservation Section the chance to link conservation more strongly to the scientific work going on within UNESCO. Like Nicholson and the IBP/CT, UNESCO also emphasized the importance of integrated surveys and maps of natural resources on land and water, and the need for international cooperation when it came to the use of natural resources. While IBP was in full swing, this focus on international cooperation in resource research led to what came to be called the Biosphere Conference in 1968, which was organized by UNESCO scientists with extensive support from IUCN and IBP members.[68] In 1970, in the aftermath of the conference, first discussions took place about a Man and the Biosphere Program (MAB), a counterpart to the IBP with a focus on the social impact of resource management and an emphasis on the conservation of local specificities.[69] The MAB, which UNESCO scientists hoped could take over the IBP's results after 1974, offered an outlook on a new and financially attractive platform for conservationists and their advice after the IBP had ended. This was also recognized within the chronically underfunded IUCN. A second group of IUCN conservationists were now hoping that their many colleagues working for IBP, after the program's potential integration into MAB, could facilitate a closer alliance between IUCN and the resource-rich UNESCO.[70] This second group differed from the conservationists around Nicholson in the question of independent, technocratic implementation of conservation as promoted by the members of the IBP/CT section. Rather than focusing on ecological processes and on calculating and managing potential productivity, as suggested by Nicholson, this second group aimed at the local conservation of ecosystems to protect the endangered species or populations they contained.

In the late 1960s, a group of former UNESCO biologists had found its way in at the top end of IUCN. This group, which included IUCN's director general Gerardo Budowski and the Union's senior ecologist Raymond Dasmann, lobbied for a stronger affiliation of the IBP/CT's work with research on natural resources conducted by UNESCO. Both Budowski and Dasmann viewed Nicholson's attempts to turn the IBP/CT into a prototype for global, conservation-steered ecosystem management with some skepticism.[71] Their critique was exemplified by the classification system they used to oppose that of Fosberg. Both Budowski and Dasmann were immediately critical of the Fosberg classification, as well as the check-sheet approach favored by Nicholson.

In 1970, Budowksi, IUCN's new director general, invited his former UNESCO colleague Dasmann to join his staff.[72] Dasmann—an American zoologist who had been involved in the drafting of the Biosphere Conference background papers—joined IUCN as the new senior ecologist. As such, he was responsible for IUCN's scientific management during Budowski's frequent periods of absence. A look at the professional development of Dasmann helps one understand the origins of this criticism. Dasmann's pathway to conservation had been quite different from that of Nicholson. For one, Dasmann pursued it in a different sociopolitical environment. For another, unlike Nicholson, Dasmann had an academic background in biology and had already gathered significant experience in the field during his student years. Dasmann had studied zoology at Berkeley under Aldo Leopold's son Starker in the postwar period, and graduated on a study of deer populations in California in the mid-1950s.[73] From the start, he was fascinated by ecological approaches that treated animal species as part of a larger ecosystem.[74] In contrast to Nicholson, however, Dasmann did not aim to explore the productivity of natural and managed ecosystems. Rather, according to Dasmann, conservation work was to focus on the management of particular species populations in those ecosystems that naturally formed their environment.[75]

Another point in which Dasmann's conservation approach differed from Nicholson's was the role it assigned to conservation experts. Dasmann did not support the technocratic and centralist management role for conservation experts assumed by Nicholson. Nicholson's experiences as a student and early professional in wartime England had seemed to confirm the advantages of centralized scientific planning.[76] Dasmann, fifteen years Nicholson's junior, spent his student years at Berkeley reading Lewis Mumford's techno-pessimistic sociological writings and William Vogt's *Road to Survival*—works that propagated decentralized, individual action and local resource management approaches.[77] He also studied Aldo Leopold's *A Sand County Almanac*, which very much based its valorization of nature on human ethics rather than subordinating human society to an undifferentiated utilitarian economy.[78] This way of thinking fed into Dasmann's postgraduate work at Minnesota University and Humboldt University in California, where he taught resource management in close cooperation with the U.S. Forest Service and the Fish and Wildlife Service.[79] Like Nicholson, Dasmann was not a fan of strictly sanctuary-based conservation approaches that closed off nature from all human influences. Yet Dasmann's idea of environmental management did not include human societies or populations as entities to be managed as rigidly as Nicholson suggested. According to Dasmann, it was the conservationists' responsibility to see to the careful handling and managing of the planet's resources in order to guarantee human wellbeing, not to manage human conduct.[80]

Dasmann also had a strong concern for the local consequences of conservation measures. His experiences during his Ph.D. fieldwork in Southern Rhodesia (now Zimbabwe) had made him skeptical of an undifferentiated application of global conservation principles. While working on a wildlife management and game ranching project in Southern Rhodesia in 1963, Dasmann witnessed the local opposition to ranching rates, which were set by Northern researchers and conservationists, in places with hunting traditions and where local administrations were not involved in implementing the cropping limits.[81]

Dasmann brought these experiences with him to IUCN, where he promoted a version of conservation and development that was quite different from the one Nicholson had in mind.

While Nicholson thus promoted the ideology of the IBP, Dasmann preferred an approach that was much closer to that of UNESCO. As explained, IBP/CT promoted global ecosystem studies that focused on potential productivity. It aimed at recording separate data units, which would allow for supposedly objective computer surveys for productive land management. UNESCO's classification was more descriptive. It was neither suited for nor aimed at computer calculations. Instead it had been developed for the mapping of resources for human use or preservation. It was not to be used for a globally comparative survey, but to serve for the creation of sets of local maps for particular conservation or development projects.

UNESCO's vegetation classification was published in 1973 by the UNESCO Committee on Classification and Mapping of the World's Vegetation. It was based on a list supplied by the German botanists Josef Smithüsen and Heinz Ellenberg, and on further drafts by Ellenberg and Dieter Mueller-Dombois from 1967 and 1969.[82] It provided sets of symbols, color schemes, and labels to mark vegetational zones on geographical regions or country maps with a scale of 1 to 1 million (figures 2.4 and 2.5). The architects of the mapping system did not assume that vegetational structures alone, without any indication of particular species, could be used to identify types of ecological habitats or environments.[83] Therefore, the UNESCO classification linked more structural elements, such as the spatial arrangements of plants, to particular associations of indicator species, which Ellenberg called "ökologische Gruppen" or "synusia" (figure 2.4).[84]

Unlike the Fosberg system, the UNESCO classification system was not a tool to advance basic ecological research. Instead of testing these links first, UNESCO added supplementary terms that were believed to coincide in particular geographical conditions, such as climate, soil, and landform. On the basis of these groups of species and the climatic conditions they were thought to live in, an accompanying vegetation map included a color and symbol code for 225 vegetation types (figure 2.5). In this way, endangered vegetation units in different regions could be singled out on a map for conservation or other purposes.

Figure 2.4. UNESCO's classification units, 1973. Reproduced from UNESCO, *International Classification and Mapping of Vegetation* (Paris: UNESCO, 1973).

197 V.C.5a   *Tropical alpine open to closed bunch-grass communities with a woody synusia of tuft plants (Espeletia, Lobelia, Senecio)*, microphyllous to leptophyllous dwarf-shrubs and cushion plants, often with woolly leaves. Above timberline in low latitudes: Páramo and related vegetation types without snow in the alpine regions of Kenya, Colombia, Venezuela, etc.

Figure 2.5. UNESCO's vegetation symbols, 1973. Reproduced from UNESCO, *International Classification and Mapping of Vegetation* (Paris: UNESCO, 1973).

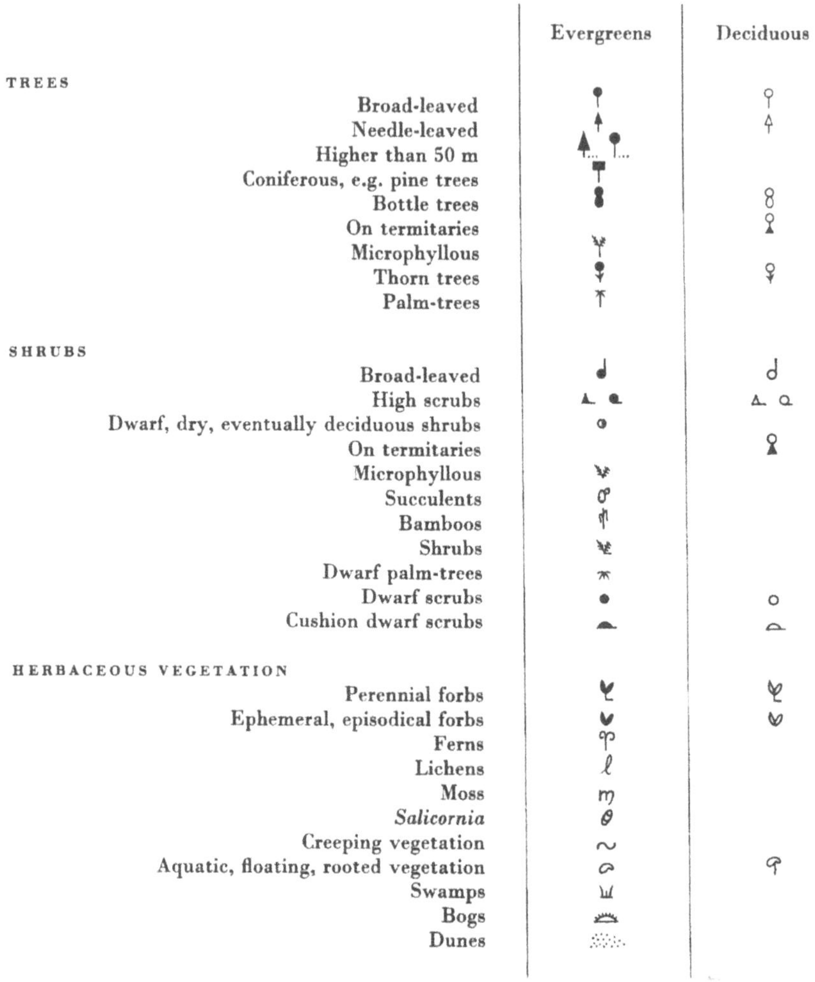

	Evergreens	Deciduous
TREES		
Broad-leaved		
Needle-leaved		
Higher than 50 m		
Coniferous, e.g. pine trees		
Bottle trees		
On termitaries		
Microphyllous		
Thorn trees		
Palm-trees		
SHRUBS		
Broad-leaved		
High scrubs		
Dwarf, dry, eventually deciduous shrubs		
On termitaries		
Microphyllous		
Succulents		
Bamboos		
Shrubs		
Dwarf palm-trees		
Dwarf scrubs		
Cushion dwarf scrubs		
HERBACEOUS VEGETATION		
Perennial forbs		
Ephemeral, episodical forbs		
Ferns		
Lichens		
Moss		
Salicornia		
Creeping vegetation		
Aquatic, floating, rooted vegetation		
Swamps		
Bogs		
Dunes		

For Senior Ecologist Dasmann and others at IUCN, UNESCO's system— combining into one category the environmental records on local soil and climate, formational data on vegetation density and distribution, and information on particular species—was preferable over IBP/CT's strict separation of recorded characteristics. Later, a synthesis volume of 1980, recollecting the IBP/CT sections achievements, explained that many conservationists recognized from the beginning that the "information derived from the use of the Fosberg classification would be insufficient for the full purposes of the check sheet survey, with its emphasis on conservation."[85] It lacked information on threatened ecosystems or species. Dasmann especially explained his discontent when corresponding and publishing on the matter. The focus of conservation, after all, was on threatened species and populations, not on the physical structures of vegetation or potential productivity.[86] He was not alone in this view. Former IUCN conservationist and IBP/CT member Duncan Poore explained retrospectively that he disagreed with approaches such as Nicholson's that focused on long-term data collection and neglected the focus on those ecosystems necessary for the conservation of endangered nature.[87] In particular, the "predictive power" that Nicholson assigned to the check-sheet survey was criticized, as it was believed to do little to solve the pressing conservation problems of the time.[88]

Alongside the focus on species, UNESCO's way of ordering nature was in two additional ways different from Fosberg's classification and Nicholson's interpretation thereof. First, the UNESCO maps were restricted to "unspoiled" vegetation. In contrast to Nicholson's project of mapping all types of managed and unmanaged ecosystems, the UNESCO classification was based on natural climax vegetation and near-climax vegetation, "not wheat fields" or "banana plantations."[89] The idea that regional plant and animal communities in isolation steadily strove toward a balanced and stable climax state had been most prominently described in the work of the American ecologists Frederic Clements and Victor Shelford.[90] Recording these different regional climax communities—so-called biomes—did not generate new knowledge that could contribute to more efficient land-use planning and management. Yet for Dasmann, UNESCO's interpretation of Clements's and Shelford's biomes presented the appropriate starting point to develop regional maps of the major habitats that might require conservation.[91]

Second, rather than imposing universally valid criteria from London or the Smithsonian Institution to be disseminated throughout all world regions, the UNESCO system was much more open to local contributions. While the IBP check sheet was centrally designed and analyzed, the UNESCO classification allowed local mapping criteria based on a combination of regionally significant floral and structural factors. Local terminology could also be taken into account: "Locally established terms meaningful to the inhabitants of the respective region (e.g., campo cerrado) may be added. . . . In this manner vegetation maps are meaningful to local users as well as to a world-wide audience. This is especially important for comparative studies."[92]

With regard to the two classifications, we thus see very different politics of global conservation at work. UNESCO's mapping system reflected its members' take on the scope of conservation expertise and its potential implementation, which had—unlike Nicholson's doctrinal universalism—always allowed for regional approaches and initiatives, an approach that aligned well with Dasmann's own experiences. Critical of all forms of centralized technocracy, he considered undifferentiated top-down approaches unworkable when the diverse values and priorities of local communities, regions, or nation-states were at stake.[93] In a publication called *Environmental Conservation*, which was aimed at defining conservation as a scientific discipline, Dasmann proclaimed, "No single piece of legislation or governmental reorganization will guarantee that environmental conservation will become a reality. [Environmental struggles] will be found at city, county, state, national, and international levels."[94] Neither in conservation, nor in development issues should experts "impose something from without," as Dasmann continued to stress.[95]

In contrast to the IBP ecologists, Dasmann and those favoring closer links between IUCN and UNESCO not only stressed the common duties of its member states to carefully gauge projects of resource use and conservation, but also pointed at the countries' own rights to locally determine development rates, allowing for more bottom-up conservation incentives.[96] IBP/CT's cybernetics was supposed to provide an objective analysis of the biosphere-technosphere-nexus, allowing for centralized and top-down management measures that were universally applicable and value free. In contrast, Dasmann saw society and man as the determining factors, both for the organization and the content of conservation. Moreover, Dasmann came to doubt that exactly the same universal rules for conservation could apply everywhere on the globe. From the mid-1970s onward, he propagated a form of "bioregionalism" that linked nature protection to local traditions and cultures.[97] Measurements should not be "filtered from the top," but designed and implemented "at grass root level," respecting the "importance of indigenous peoples and their cultures."[98] Dasmann's vision for the science, organization, and implementation of conservation advice was thus not only irreconcilable with Nicholson's central technocratic planning, but was much more inclusive of local exceptions, priorities, and traditions, and put a lot of emphasis on fieldwork and a diversity of methods, as well as local experiences and governance.

In line with Dasmann's way of thinking, a number of IUCN members were in favor of integrating the results of the IBP more closely into the UNESCO program to allow for a continuation of IBP efforts after the end of its ten-year period. At an interagency consultation meeting at UNESCO in 1969, it was decided that the "structures created through IBP, the international networks amongst scientists, need[ed] to be kept alive" and should be integrated into UNESCO's Man and the Biosphere Program after the closure of the IBP's field programs around 1972.[99] In 1971, an International Cooperation Council was formed with representatives from UNESCO, IUCN, and the IBP to facilitate this process. It was to ensure that the follow-up of the IBP entailed a stronger alignment with UNESCO's focal areas, the study of particular ecosystem types and climatic regions in particular locations, as well as a much stronger human-centered approach to conservation.

A New Global Approach for Conservation

With the decision to continue the IBP/CT's project within UNESCO, a clear choice was made regarding the organizational future of conservation. At first glance, it seemed as if the IUCN expert group around Dasmann, promoting an alliance with UNESCO's Rational Use of Natural Resources Program, largely prevailed over Nicholson's attempts to use the IBP's Conservation Section for his technocratic reform plans. In some ways, the ecological approach followed within the MAB was significantly different from the encompassing systems approach promoted by the IBP.[100] Proposals by the U.S. national IBP Committee for a Program for the Analysis of World Ecosystems, to counterbalance MAB's work on particular regions, were steadfastly rejected.[101] It would be wrong to conclude, however, that the IBP/CT's work remained unrecognized by the future members of the MAB. Despite the divergence from Nicholson's original plans, the IBP/CT contributed to the conceptual framework of the MAB to a substantial degree. In fact, after the IBP, a new approach to conservation advice emerged among conservationists involved in the work of IUCN and UNESCO's MAB that combined aspects from both approaches.

After the closure of the IBP, this double intellectual legacy was noticeable within the conservation discourse of both IUCN and UNESCO. From around 1970 onward, documents and discussions at both organizations were studded with references to ecosystems and their relevance for conservation practices as both Nicholson and Dasmann had used them.[102] When in 1971 the IUCN took over the compilation of the IBP/CT's check-sheet survey, IUCN's commissions, which previously had often followed quite diverse agendas, came to agree on the usefulness of such a study and adopted a loosely defined ecosystem approach for their programs.[103] With this came another important change. From about the same time onward, IUCN documents contained a renewed emphasis on the potentially global scope of conservation work that went beyond the focus on particularly threatened ecosystems in particular geographic regions. At IUCN's General Assembly of 1972, Budowski, Dasmann, and Frank Nicholls made the development of a world network of reserves for science-based conservation a major program point for IUCN for the coming years. Echoing the IBP/CT's early agenda, this network was to include representative ecosystem samples "in addition to the current *ad hoc* approach."[104] It moreover constituted a main point for cooperation with UNESCO, where similar plans were formulated for the MAB.

In the following years, a joint expert panel of IUCN conservationists, including former IBP members, and scientists of UNESCO worked to prepare the follow-up of the IBP. In 1973, a task force was created to draw up a list of criteria and guidelines for the establishment of a global reserve network under the auspices of the MAB.[105] Throughout the report of the expert panel, one finds traces of the IBP/CT's conceptual framework. In 1972, the IBP/CT, together with IUCN, had begun the evaluation of the check sheets.[106] In particular, the former IBP members among the panelists, such as Nicholson's colleague Geoffrey Radford and Worthington of the IBP's central office, made sure that the results of Nicholson's check-sheet survey were taken into account.[107] The number of returned check sheets had been smaller than expected; nevertheless, for the first time a global ecological survey had been conducted.[108] The results were collected by Radford for IUCN at the Nature Conservancy's research station at Monks Wood. Paired with a respective geo-code that revealed the geographical origins of the results, the data resulting from the survey had been stored first on punch cards and later on magnetic tapes to be used with early ATLAS 2 computers.[109] This computer system, then, provided responses to inquiries on the characteristics of the vegetation at particular sites, on the location of particular vegetation structures, and on additional climatic, species, and conservation-related information that had been recorded.[110] In the end, the task force of IUCN and UNESCO scientists recommended integrating the results of the IBP/CT survey with UNESCO's resource maps.[111] Based on this, the panel's final report made ecosystem-based "biosphere reserves" the main tool for conservation, even if this previously had not been part of UNESCO's research programs on natural resources.[112]

Following the panel's recommendations, in 1975 the Hungarian IUCN biologist and biogeographer Miklos Udvardy published yet another classification on behalf of UNESCO. Udvardy's classification system, which forms the decision framework for UNESCO biosphere reserve nominations still today, certainly benefitted from the trials and errors of the IBP check-sheet survey.[113] The original authors of both the UNESCO and the IBP/CT classification, Ellenberg and Fosberg, contributed to the drafting of Udvardy's system. They took over parts of the basic structure of UNESCO's classification, but subordinated Dasmann's biomes to larger biogeographical regions. These differed in their geographical location, yet it was assumed their ecosystems could be regarded as quite similar in composition and structure. This, then, resembled the rationale behind Fosberg's ecosystem classification, while altogether the worldwide system of biogeographical regions constituted the global framework to which Nicholson had aspired with his IBP check-sheet survey.

Moreover, a similar blend of Dasmann's and Nicholson's approach can be found in the proclaimed objective of the MAB conservation projects. For the conservation of ecosystems in the different biogeographical regions, information on fauna and flora was deemed essential, similar to what Dasmann had demanded. Yet, the protection of endangered species was only one of the aims behind the classification. In addition, ecosystem research—now recognized, in the spirit of Nicholson, as the "necessary basis for further development of the life sciences"—was deemed a goal equally important.[114] On top of this, the concept of MAB biosphere reserves extended the focus of UNESCO's resource studies from the purely natural to the human-influenced, built environment in line with the IBP and Nicholson's British experiences.[115]

Alongside the new focus on ecologically based conservation of ecosystems in different biogeographical regions, the integration of the IBP's results in the conservation framework of UNESCO also brought about changes regarding the type and dissemination of conservation expertise. Although conservationists involved in the MAB program did not maintain the technocratic universalism that Nicholson had linked to the protection of ecosystems, neither did they rely on the original UNESCO approach of local resource mapping alone. In fact, again aspects of both Dasmann's and Nicholson's approach to internationally organized conservation work played into the implementation framework of the MAB.

On the one hand, the endeavors of conservationists engaged in the MAB were supposed to be problem oriented, placing priority on those world regions where living resources were at particular risk and where their supply for local human populations was unsecured. In general, within postwar international organizations such as the UN, discussions on the scientific management of natural resources became more and more closely linked to questions related to the economic development of countries in the global South.[116] This led to studies on particular ecosystems in areas prone to droughts or floods, typically found in sub-Saharan Africa and Southeast Asia. The MAB in fact continued and extended the regional focus of both UNESCO's existing research in developing arid and tropical zones and IUCN's special area programs related to nature reserves and the protection of threatened species in the recently decolonized global South. In particular, this corresponded to Dasmann's call for the adaptation of conservation work to each region's societal and ecological situation.[117] On the other hand, however, the MAB offered to conservationists a new infrastructure that could potentially bridge regional and global environmental concerns, as had been envisioned by the members of the IBP/CT section. The MAB, after all, presented a global framework into which local conservation priorities and projects were integrated. Moreover, the strengthened links with the UN system promised especially for IUCN scientists new ways of spreading their conservation advice. Rather than subordinating conservation to local politics, working as a scientific advisory body to the UN and its agencies seemed to open up new potential consultative functions for IUCN conservationists, related to Nicholson's ideas on an independent advisory body in conservation matters.[118]

In general, therefore, the IBP brought about a twofold legacy for international and science-based conservation, pertaining to both the intellectual foundations of conservation work and the institutional organization of international conservation. After the IBP, conservationists engaged in IUCN and

UNESCO pursued a type of ecosystem conservation that was concerned with both natural and modified systems, and that looked to both regional priorities and global standards. Moreover, ideas on both top-down planning and stand-alone regional projects made way for new forms of scientific cooperation between scientists at IUCN and intergovernmental agencies of the UN, such as UNESCO. Both aspects of this new conservation approach would play an important role in conservation cooperation between IUCN and UNESCO in the years to come, and, to a significant degree, shape the understanding of conservation science, methods, and expert roles held by the members of IUCN.

Conclusion

This chapter has looked at the reorganization of the scientific and international organization of global nature conservation through the involvement of IUCN conservationists in the projects of the IBP. From the beginning of the IBP in 1964, IBP/CT convener Max Nicholson attempted to reform international conservation by strengthening and consolidating its scientific basis and by broadening its sphere of competence to scientifically planned land-use practices. After 1970, a second group of conservationists around IUCN's senior ecologist Ray Dasmann tried to use the IBP to develop a closer cooperation with UNESCO's work on regional resource maps. Both groups saw their view on conservation as global and science based. For each group, however, this entailed very different ideas on the objectives, practices, and scales associated with conservation work and expertise.

The discrepancies between the two groups regarding their political ideas, organizational cultures, and ecological approaches did not lead to the definite exclusion of one approach by the other. Rather than continuing a single tradition, ecologists at IUCN and UNESCO took up elements of both conceptions. Under the influence of the IBP, the concept of ecosystems—holistic ecological units that encompassed different species populations and their shared nonliving environment—had entered most conservation discourses. While biogeographical ideas on mapping communities in their environment did not fade, the idea of ecosystems as the unit on which to focus opened up the conservation discourse both in terms of global scope and responsibility for the living environment and living natural resources at large. After the IBP, conservationists engaged in projects at IUCN and UNESCO based their work on the conservation of natural and modified ecosystems in different biogeographical regions.[119] Moreover, in the framework of the MAB, new forms of scientific cooperation also emerged between scientists at IUCN and UNESCO and seemed to promise IUCN conservationists new roles as scientific advisors to conservation-related projects in the UN.

When the UN announced their plans for the first international Conference on the Human Environment, to be held in 1972 in Stockholm, the new cross-organizational approach to ecosystem conservation seemed to put IUCN conservationists in a strong position. In the emerging context of environmental politics, planning for the planet, as proposed by Nicholson and his IBP/CT section, seemed all the more important. Within the political realm of the UN, however, such a role was not self-evident. The controversies emerging in this new context of environmental diplomacy will be the focus of chapter 3.

Notes

1. Part of this chapter has been published as an article in the *Journal of the History of Biology* under the terms of the Creative Commons Attribution 4.0 International License: https://link.springer.com/article/10.1007/s10739-015-9433-4. Simone Schleper, "Conservation Compromises: The MAB and the Legacy of the International Biological Program, 1964–1974," *Journal of the History of Biology* 50, no. 1 (2017).

2. IUCN, ed., *Proceedings of the Eighth General Assembly, Kenya 1963* (Morges: IUCN, 1964), 127.

3. E. Barton Worthington, *The Ecological Century: A Personal Appraisal* (Oxford: Clarendon Press, 1983), 140. IBP initiators were the American physicist Lloyd Berkner and the British biochemist Sir Rudolph Peters, at that time the past and the present president of the ICSU, respectively, and the Italian geneticist Giuseppe Montalenti, president of the IUBS. The variety of professions indicates the unconsolidated state of transition that biology as a discipline was in at the time.

4. In total, the IBP consisted of seven subsections, two others on the productivity of terrestrial communities and plants, a section each on freshwater and marine productivity, a section on human adaptability, and a final, more practically oriented section on applied biology.

5. John Krige and Kai-Henrik Barth, eds., *Global Power Knowledge: Science and Technology in International Affairs* (Chicago: University of Chicago Press, 2006); Helmuth Trischler and Hans Weinberger, "Engineering Europe: Big Technologies and Military Systems in the Making of 20th Century Europe," *History and Technology* 21, no. 1 (2005); David C. Coleman, *Big Ecology: The Emergence of Ecosystem Science* (Berkeley: University of California Press, 2010), 2; David G. Raffaelli and Christopher L. Frid, eds., *Ecosystem Ecology: A New Synthesis* (Cambridge: Cambridge University Press, 2010); E. Barton Worthington, "IBP: International Goals," *Science* 161, no. 3839 (1968). For different forms of scientific internationalism and international science in the twentieth century, see Elisabeth Crawford, *Nationalism and Internationalism in Science, 1880–1939: Four Studies of the Nobel Population* (Cambridge: Cambridge University Press, 2002); Elisabeth Crawford, Terry Shinn, and Sverker Sörlin, eds., *Denationalizing Science: The Contexts of International Scientific Practice* (Dordrecht: Springer Science and Business Media, 2013); Frank Greenaway, *Science International: A History of the International Council of Scientific Unions* (Cambridge: Cambridge University Press, 2006); Geert J. Somsen, "A History of Universalism: Conceptions of the Internationality of Science from the Enlightenment to the Cold War," *Minerva* 46, no. 3 (2008). For accounts on

technocratic planning, see James Ferguson, *The Anti-Politics Machine: "Development,"* *Depoliticization, and Bureaucratic Power in Lesotho* (Minneapolis: University of Minnesota Press, 1990); Tim Mitchell, *Rule of Experts: Egypt, Techno-Politics, Modernity* (Berkeley: University of California Press, 2002).

6. Coleman, Golley and Kwa have identified the IBP as the moment when ecosystem ecology received widespread international support and recognition by scientists and policymakers as a relevant discipline in dealing with the environmental problems observed at the time. Coleman, *Big Ecology*; Frank B. Golley, *A History of the Ecosystem Concept in Ecology: More Than the Sum of the Parts* (New Haven: Yale University Press, 1993); Chunglin Kwa, "Representations of Nature Mediating between Ecology and Science Policy: The Case of the International Biological Programme," *Social Studies of Science* 17, no. 3 (1987): 163; John McCormick, *Reclaiming Paradise: The Global Environmental Movement* (Bloomington: Indiana University Press, 1991).

7. Waddington in E. Barton Worthington, *The Evolution of IBP* (Cambridge: Cambridge University Press, 1975), 4.

8. Robert B. Boardman, *International Organization and the Conservation of Nature* (Bloomington: Indiana University Press, 1981), 51.

9. Peter J. Bowler and Iwan Rhys Morus, *Making Modern Science: A Historical Survey* (Chicago: University of Chicago Press, 2005), 225; Robert McIntosh, *The Background of Ecology: Concept and Theory* (Cambridge: Cambridge University Press, 1986), 108.

10. IUCN, ed., *Proceedings of the Fifth General Assembly: Edinburgh, June 1956* (London: Society for the Promotion of Nature Reserves in collaboration with the Nature Conservancy for the IUCN, 1957), 45ff.; also see Martin W. Holdgate, *The Green Web: A Union for World Conservation* (Gland: IUCN, 1999), 63.

11. E. Max Nicholson, "Research and Natural Areas," in *First World Conference on National Parks: Proceedings of a Conference, Seattle, Washington, June 30–July 7, 1962*, ed. Alexander B. Adams, IUCN, and UNESCO (Washington, DC: National Park Service, U.S. Department of the Interior, 1964), 90–94.

12. McIntosh, *The Background of Ecology*, 227; Raffaelli and Frid, *Ecosystem Ecology*, 7–10.

13. Kenneth J. Collins and Joseph S. Weiner, *Human Adaptability: A History and Compendium of Research in the International Biological Programme* (London: Taylor & Francis, 1977), cited in Raffaelli and Frid, *Ecosystem Ecology*, 5; Worthington, *The Evolution of IBP*.

14. Anya Plutynski, "Ecology and the Environment," in *The Oxford Handbook of the Philosophy of Biology*, ed. Michael Ruse (Oxford: Oxford University Press, 2008), 511; also see Timothy Farnham, "A Confluence of Values: Historical Roots of Concern for Biological Diversity," in *The Routledge Handbook of Philosophy of Biodiversity*, ed. Justin Garson, Anya Plutynski, and Sahotra Sarkar (London: Routledge, 2016), 17ff.

15. John Krige and Kai-Henrik Barth, "Science, Technology, and International Affairs: New Perspectives," in *Global Power Knowledge*, ed. Krige and Barth, 14; see, e.g., IUCN, ed., *Proceedings of the Seventh General Assembly, Warsaw 1960* (Morges: IUCN, 1960), 97.

16. Stanley Johnson, *UNEP: The First 40 Years; A Narrative* (Nairobi: UNON/Publishing Section Service, 2012), 234–35.

17. Worthington, *The Ecological Century*, 165.

18. Johnson, *UNEP: The First 40 Years*, 228.

19. IUCN, *Proceedings of the Eighth General Assembly, Kenya 1963*, 63, 78.

20. E.g., IUCN, "General Statement (Revised): IUCN's African Special Project (ASP) 1960–1963," 1961, LSA EMN/IBP: Box 8, Folder "CCTA/IUCN Arusha 5–12 September 1961"; also see Rachelle Adam, *Elephant Treaties: The Colonial Legacy of the Biodiversity Crisis* (Lebanon, NH: University Press of New England, 2014); Helen Tilley, *Africa as a Living Laboratory: Empire, Development, and the Problem of Scientific Knowledge, 1870–1950* (Chicago: University of Chicago Press, 2011).

21. August W. Küchler, Jorge M. Montoya Maquin, and UNESCO, *The UNESCO Classification of Vegetation: Some Tests in the Tropics* (Paris: UNESCO, 1970); also see UNESCO, *Sixty Years of Science at UNESCO, 1945–2005* (Paris: UNESCO, 2006).

22. IUCN, *Proceedings of the Eighth General Assembly, Kenya 1963*, 132.

23. For an impressively detailed and extremely well executed history of science-based nature conservation in South Africa, see Jane Carruthers, *National Park Science: A Century of Research in South Africa* (Cambridge: Cambridge University Press, 2017). For the South African context of international nature conservation in the postwar period, see Carruthers, *National Park Science*, 150–155; J. P. Singh, *United Nations Educational, Scientific, and Cultural Organization (UNESCO): Creating Norms for a Complex World* (London: Routledge, 2010). Some historians have claimed that scientific internationalism within the UN system was in fact often "sacrificed on the altar of nationalism and patriotism." Krige and Barth, "Science, Technology, and International Affairs," 13.

24. Anna-Katharina Wöbse, "'The World after All Was One': The International Environmental Network of UNESCO and IUPN, 1945–1950," *Contemporary European History* 20, no. 3 (2011): 340.

25. A membership in ICSU was dependent on national academies to finance the membership fees themselves. This changed slightly with funds sustained by UNESCO in the 1960s, which allowed matching of less affluent scientific communities' membership fees. Nevertheless, the members of the academies usually came from the scientific elite.

26. Somsen, "A History of Universalism," 365–70; Greenaway, *Science International*, 86–89.

27. Whereas the launch of Sputnik contributed to growing tensions between USSR and U.S. diplomacy, within IUCN, the satellite was seen as a contribution to the world scientific effort and a potential new technology for truly global planning. For a related account of Sputnik and the arms race, see, e.g., Walter Hixson, *American Foreign Relations: A New Diplomatic History* (New York: Taylor & Francis, 2015), 277ff. ICSU's take on scientific internationalism is captured in Greenaway, *Science International*, 91.

28. For an account of socialist countries in ICSU's IBP, see Doubravka Olsakova, "The International Biological Program in Eastern Europe: Science Diplomacy, Comecon and the Beginning of Ecology in Czechoslovakia," *Environment and History* 24, no. 4 (2018).

29. Within the IBP at large, emphasis was put on the nongovernmental basis of its membership. The international subcommittees, consisting of scientific experts who were appointed for their competencies, were subject to the central executive board, the Special Committee (SCIBP). Additionally, there were national committees of participating national scientific communities, but these were based on the themes and the programs of the international subsections. Nicholson presented his subcommittee at the second IBP General Assembly in Paris in 1965. All members had some affiliation with IUCN. Among others, the section comprised the British Duncan Poore, Arthur R. Clapham, Geoffrey Radford, and George Peterken, Nicholson's colleagues from the Nature Conservancy, and the Antarctic scientist Martin Holdgate. Moreover, with Lee Talbot, who

had been IUCN's first staff ecologist, the Smithsonian ecologists Edward Graham and Raymond Fosberg, and the German-American botanist Dieter Mueller-Dombois, the CT section also had a strong American influence. Other well-known names included the Soviet zoologist Andrei Bannikov, the Polish geologist Walery Goetel and botanist Anna Medwecka-Kornas, the Ecuadorian botanist Misael Acosta Solis, and the French naturalist Jean Dorst. The Egyptian botanist Mohamed Kassas joined a few years later. Most of them had studied at renowned European or American universities, had significant field experience, and could draw on extensive national and international research networks. Thus they could provide the IBP/CT in London with the access to the relevant science communities and regions that was needed to run an international program as poorly funded as the IBP (E. Max Nicholson, "IBP Section CT / Conservation of Terrestrial Communities: Progress Report for the Sectional Committee," 1965, RSA SCIBP: NHM, Box 1). Nicholson later recalled how the IBP/CT could "exploit" IUCN's scientific network and could "work through established conservation bodies in certain countries," such as the Nature Conservancy in Great Britain; Nicholson cited in Worthington, *The Evolution of IBP*, 12–13.

30. Early examples of such reasoning are conservation attempts in the Engadin Valley in Switzerland around 1910 and Albert National Park in the then Belgian Congo in the 1920s; Patrick Kupper, *Creating Wilderness: A Transnational History of the Swiss National Park* (New York: Berghahn Books, 2014), 85; Bernhard Gissibl, Sabine Höhler, and Patrick Kupper, *Civilizing Nature: National Parks in Global Historical Perspective* (New York: Berghahn Books, 2012), 212.

31. Edward Graham, "A New Network of Reserves," *New Scientist*, 14 October 1965, 127.

32. Ibid., 128.

33. Elena Aronova, Karen Baker, and Naomi Oreskes, "Big Science and Big Data in Biology: From the International Geophysical Year through the International Biological Program to the Long Term Ecological Research (LTER) Network, 1957–Present," *Historical Studies in the Natural Sciences* 40, no. 2 (2010): 208.

34. E. Max Nicholson to Jean George Baer, 15 June 1965, LSA EMN/IBP: Box 2, Folder "Relation with IUCN 1963–1972."

35. E.g., W. Frank Blair, *Big Biology: The US/IBP* (Stroudsburg: Dowden, Hutchinson & Ross, 1977); Worthington, "IBP: International Goals."

36. E.g., McCormick, *Reclaiming Paradise*; Robert Boote, "Obituary. Max Nicholson: The Prime Mover of the Nature Conservancy and the World Wildlife Fund, Who Helped Inspire Nature Reserves and Ecological Research," *The Guardian*, 28 April 2003, accessed 24 March 2019, http://www.theguardian.com/news/2003/apr/28/guardianobituaries.highereducation; E. Max Nicholson, "Unpublished Autobiography," n.d., ca. 2000–2003, AL EMN: Boxes A.7–A.12.

37. Gregory J. Cooper, *The Science of the Struggle for Existence: On the Foundations of Ecology* (Cambridge: Cambridge University Press, 2007), 46.

38. Peder Anker, *Imperial Ecology: Environmental Order in the British Empire, 1895–1945* (Cambridge, MA: Harvard University Press, 2001); Sharon E. Kingsland, *The Evolution of American Ecology, 1890–2000* (Baltimore: Johns Hopkins Press, 2005), 240ff. The environmental historian Peder Anker has described how in the 1920s and 1930s Tansley and Huxley, at the time good friends of Nicholson, belonged to a particular Oxford school of human ecologists. In ecology applied to human society, this group saw the

solution to the looming economic depression in Britain and other parts of the Western world; Julian Huxley, *The Stream of Life* (London: Watts, 1926); Julian Huxley et al., *Scientific Research and Social Needs* (London: Watts & Co., 1934); Arthur G. Tansley, *The Values of Science to Humanity: The Herbert Spencer Lecture, Oxford University, 2 June 1942* (London: G. Allen & Unwin, 1942); Anker, *Imperial Ecology*, 221.

39. For an account of early ecology-based nature protection and research in Soviet reserves, see Douglas R. Weiner, "The Changing Face of Soviet Conservation," in *The Ends of the Earth: Perspectives on Modern Environmental History*, ed. Donald Worster (Cambridge: Cambridge University Press, 1988). For systematic and centrally planned nature management in the Soviet Union during the Cold War period, see Doubravka Olsakova, ed., *In the Name of the Great Work: Stalin's Plan for the Transformation of Nature and Its Impact in Eastern Europe* (New York: Berghahn Books, 2016).

40. Nicholson's organizational talent granted him a post in the deputy prime minister's office in 1945 and a seat in the official Advisory Council on Scientific Policy in 1948. In the postwar years, he contributed to the implementation of legislation such as the 1947 Town and Country Planning Act and the 1949 National Parks and Access to the Countryside Act; Boote, "Obituary: Max Nicholson." These acts established that "physical planning should be conceived as a national, rather than a local, responsibility," and they transferred the rights to maintain and manage the British National Parks to a national research committee (Town and Country Planning Act, 1947 [10 & 11 Geo. VI c. 51], 1948). In 1952, Nicholson became director general of this committee, which by then had become part of the British Nature Conservancy; Stephen Bocking, *Ecologists and Environmental Politics: A History of Contemporary Ecology* (New Haven: Yale University Press, 1997). For a biography of Nicholson and a comprehensive analysis of how his political views influenced his technocratic ideas on the management and conservation of natural resources, see Mark Toogood, "Beyond 'the Toad beneath the Harrow': Geographies of Ecological Science, 1959–1965," *Journal of Historical Geography* 34, no. 1 (2008).

41. Arthur G. Tansley, *The British Islands and Their Vegetation* (Cambridge: Cambridge University Press, 1939).

42. E. Max Nicholson, *Handbook to the Conservation Section of the International Biological Programme* (London: IBP/CT, 1968); *The Environmental Revolution: A Guide for the New Masters of the World* (London: Penguin Books, 1970), 64.

43. Nicholson, *The Environmental Revolution*, 286.

44. E. Max Nicholson, *Britain's Nature Reserves* (London: Country Life Ltd., 1957); also see E. Max Nicholson, "Rural Landscape as a Habitat for Flora and Fauna in Densely Populated Countries," an IUCN working paper from 1952 (IUCNL).

45. IUCN, *Proceedings of the Fifth General Assembly: Edinburgh, June 1956*, 82–85.

46. Nicholson, *The Environmental Revolution*, 129; *The Big Change: After the Environmental Revolution* (New York: McGraw-Hill, 1973). Chairing IUCN's Fifth General Assembly in Edinburgh in 1956, Nicholson criticized the decision-making capacities of the Governing Council: "La véritable U.I.C.N. est l'émanation invisible de tous ceux qui demeurent attachés aux idées de la conservation et à l'écologie qui ne peut subsister si elle n'est pas renfermée dans un corps sain" (Nicholson, cited in IUCN, *Proceedings of the Fifth General Assembly: Edinburgh, June 1956*, 50). In 1970, he expressed dissatisfaction with the lack of inclusion of the ecological sciences in international nature conserva-

tion: "At world level, all previous efforts by the International Union of Biological Sciences and the International Union for the Conservation of Nature to interest ecologists in giving the necessary minimum support for an international working group of their own have come to nothing" (Nicholson, *The Environmental Revolution*, 289).

47. Systems science set out to study biological systems "in which everything affects everything else." One of the earliest systems ecologists, Kenneth Watt, defined a system as "an interlocking complex of processes characterized by many reciprocal cause-effect pathways." Kenneth E. Watt, *Systems Analysis in Ecology* (New York: Academic Press, 1966), 2.

48. E. Max Nicholson, *Conservation and the Next Renaissance: The Horace M. Albright Conservation Lectureship* (Berkeley: University of California, Berkeley, School of Forestry, 1964).

49. Debora Hammond, *The Science of Synthesis: Exploring the Social Implications of General Systems Theory* (Boulder: University Press of Colorado, 2003); Eugene P. Odum, *Ecology* (New York: Holt, Rinehart and Winston, 1963); Kenneth E. Watt, "Use of Mathematics in Population Ecology," *Annual Review of Entomology* 7, no. 1 (1962): 253. Historians have aptly described how in the age of computer technology new ideas of system thinking merged with cybernetic calculations; see Bowler and Morus, *Making Modern Science*; McIntosh, *The Background of Ecology*; Lars Skyttner, *General Systems Theory: Problems, Perspectives, Practice* (London: World Scientific Publishing, 2005).

50. Howard T. Odum, *IBP Symposium: Environmental Photosynthesis* (Washington, DC: American Association for the Advancement of Science, 1967). Cyberneticists at the time predicted the eventual disintegration of traditional disciplinary boundaries under one shared set of systems theories; Ronald R. Kline, *The Cybernetics Moment: Or Why We Call Our Age the Information Age* (Baltimore: Johns Hopkins University Press, 2015).

51. Nicholson, *The Environmental Revolution*, 59.

52. According to Keller and Golley, energy models by, for instance, Howard and Eugene Odum played into the holistic systems thinking behind IBP in general. David R. Keller and Frank B. Golley, *The Philosophy of Ecology: From Science to Synthesis* (Athens, GA: University of Georgia Press, 2000), 208ff. Although not implying computation, these models emerged from the same conceptual framework as mathematical system models that calculated energy and nutritious cycles within closed ecosystems. The Odum brothers also devised models for different ecosystem types, independent of the geographical location, similar to the classification that Nicholson would develop.

53. E. Max Nicholson, "Environment," paper for the Columbia University Conference on International Economic Development, 20–21 February 1970, 1. LSA EMN/IBP: Box 4, Folder "World Bank etc."

54. Nicholson, *The Environmental Revolution*, 58–59.

55. Stephen Bocking, "Conserving Nature and Building a Science: British Ecologists and the Origins of the Nature Conservancy," in *Science and Nature: Essays in the History of the Environmental Sciences*, vol. 8, ed. Michael Shortland (Oxford: British Society for the History of Science, 1993), 89–114.

56. E. Max Nicholson, "Environment," paper for the Columbia University Conference on International Economic Development, 20–21 February 1970, 1. LSA EMN/IBP: Box 4, Folder "World Bank etc."

57. E. Max Nicholson to Heinz Ellenberg, 24 January 1969, RSA SCIBP: NHM Box 13, Folder 10.

58. Nicholson, *The Environmental Revolution*, 69–70.

59. Ibid., 70.

60. Nicholson concluded, "We must be content with a classification of vegetation, on the basis that vegetation is an integrated expression of the ecosystem" ("International Selection of Areas for Reserves: Papers Related to the International Biological Programme," n.d., probably 1968, UASC IBP: MS 3162, Box 8, Folder 18, "Reports on Particular Projects or Activities"). "The eventual goal will be to translate everything on to the level of entire ecosystems" (E. Max Nicholson, "For the Record," to Gerardo Budowski, Raymond Fosberg, Duncan Poore, and Heinz Ellenberg, 14 January 1969, RSA SCIBP: NHM, Box 13, Folder 10).

61. Francis Raymond Fosberg, "A Classification of Vegetation for General Purposes," *Tropical Ecology* 2 (1961), in Arthur R. Clapham, ed., *The IBP Survey of Conservation Sites: An Experimental Study* (Cambridge: Cambridge University Press, 1980), 91; George Peterken to Ray Fosberg, 27 January 1967, RSA SCIBP: NHM, Box 12, Folder 9.

62. Clapham, *The IBP Survey of Conservation Sites*, 38.

63. E. Max Nicholson to Heinz Ellenberg, 24 January 1969, RSA SCIBP: NHM, Box 13, Folder 10; also see Clapham, *The IBP Survey of Conservation Sites*, 38: "Where vegetation units are delimited in part by environmental features, correlation of such a vegetation map to environmental maps of the same area becomes problematic as this may result in circular reasoning."

64. The British Nature Conservancy had founded the Monks Wood Experimental Station not far from London in 1963 as the first research laboratory to investigate the effect of toxic chemicals on wildlife (Boote, "Obituary: Max Nicholson," 2003).

65. E. Max Nicholson to Heinz Ellenberg, 24 January 1969, RSA SCIBP: NHM, Box 13, Folder 10; compare Cooper, *The Science of the Struggle for Existence*, 45.

66. Nicholson, *The Environmental Revolution*, 68, 73.

67. Nicholson's preface to Clapham, *The IBP Survey of Conservation Sites*, 2.

68. UNESCO, ed., *Intergovernmental Conference of Experts on the Scientific Basis for Rational Use and Conservation of the Resources of the Biosphere, Paris, 4–13 September 1968: Recommendations* (Paris: UNESCO, 1968).

69. UNESCO, *Sixty Years of Science at UNESCO, 1945–2005*, 225–28.

70. UNESCO, "Can We Keep Our Planet Habitable?," *The UNESCO Courier* 22, no. 1 (1969).

71. In 1966, the American population biologist Dasmann had been suggested to UNESCO by the American conservationist Russell Train, at the time vice president of the WWF's U.S. branch and president of the North American Conservation Foundation. At UNESCO, Dasmann was to write the background papers for the Biosphere Conference of 1968. At the conference, the takeover of the IBP by the MAB was first discussed; Raymond F. Dasmann and Randall Jarrell, *Raymond F. Dasmann: A Life in Conservation Biology* (Bloomington: Xlibris Corporation, 2000), 58. The German-Venezuelan agronomy and forestry expert Budowski had been UNESCO's program specialist for ecology and conservation between 1967 and 1970.

72. Budowski had been UNESCO's Program Specialist for Ecology and Conservation between 1967 and 1970.

73. Raymond F. Dasmann, *Called by the Wild: The Autobiography of a Conservationist* (Berkeley: University of California Press, 2002), 74.
74. Dasmann and Jarrell, *Raymond F. Dasmann: A Life in Conservation Biology*, 25.
75. For an in-depth discussion on Aldo Leopold's ecology and life, see Curt D. Meine and Wendell Berry, *Aldo Leopold: His Life and Work* (Madison: University of Wisconsin Press, 2010).
76. In particular, Nicholson's wartime involvement strongly influenced his belief in a dominant role for scientific experts in the planning and managing of environmental and social problems. E. Max Nicholson, *How Britain's Resources Are Mobilized* (Oxford: Clarendon Press, 1940).
77. Lewis Mumford, *Technics and Civilization* (New York: Harcourt, Brace, 1934); William Vogt, *Road to Survival* (Brookfield: W. Sloane Associates, 1948).
78. Aldo Leopold, *A Sand County Almanac, and Sketches Here and There* (Oxford: Oxford University Press, 1949).
79. Dasmann and Jarrell, *Raymond F. Dasmann: A Life in Conservation Biology*, 28.
80. "The goal of conservation is not a narrow world in which each person restricts his own consumption and his own activities in order to enable more people to crowd in and share in ever-diminishing resources," but "an abundant and free life in a world of opportunity." Raymond F. Dasmann, *Environmental Conservation* (New York: Wiley, 1972), 433–34.
81. In Southern Rhodesia, Dasmann had worked on a project for the British wildlife conservationist Frank Fraser Darling. Back in California in 1965, Dasmann had been approached by the American conservationist Russell Train, at the time vice president of the WWF's U.S. branch and president of the North American Conservation Foundation. Train had been gathering a group of like-minded conservationists, including Fraser Darling, Vogt, Edward Graham, and Fairfield Osborn, to support his idea of a World Heritage Foundation, which combined a human-centered view on culture and nature and which he wanted Dasmann to join. Through this group, Dasmann had joined the Conservation Foundation; it was also Train who, in 1966, had suggested Dasmann to UNESCO, for which he was supposed to write the background paper for the Biosphere Conference, where the takeover of the IBP by the MAB was first discussed. Raymond F. Dasmann, *African Game Ranching* (Oxford: Pergamon Press, 1964), 63; *Environmental Conservation*, 456; Dasmann and Jarrell, *Raymond F. Dasmann: A Life in Conservation Biology*, 33–36, 58.
82. Heinz Ellenberg and Dieter Mueller Dombois, *A Key to Raunkiaer Plant Life Forms with Revised Subdivisions* (Zurich: Stiftung Rübel, 1967); UNESCO, *Expert Panel on Project 8: Conservation of Natural Areas and of the Genetic Material They Contain; Final Report* (Paris: UNESCO, 1973).
83. UNESCO, *International Classification and Mapping of Vegetation* (Paris: UNESCO, 1973); Clapham, *The IBP Survey of Conservation Sites*, 42.
84. Clapham, *The IBP Survey of Conservation Sites*, 93; Heinz Ellenberg, *Vegetation Mitteleuropas mit den Alpen in kausaler, dynamischer und historischer Sicht* (Stuttgart: E. Ulmer, 1963). One example can be found in figure 2.4: "Tropical alpine to closed bunchgrass communities with a woody synusia of tuft plants (Espeletia, Lobelia, Sencio)."
85. Clapham, *The IBP Survey of Conservation Sites*, 67; Dasmann cited by Arthur R. Clapham to Geoffrey Radford, 24 September 1975, RSA SCIBP: NHM, Box 12, Folder 9.

86. E.g., Raymond F. Dasmann, *Environmental Conservation* (New York: Wiley, 1968), 243ff.

87. Jennifer Norman, "Conservation with Dr. Poore," email sent to Simone Schleper on 13 March 2014.

88. Raymond Fosberg to Gina Douglas, 18 January 1972, RSA SCIBP: NHM, Box 12, Folder 1.

89. UNESCO, *International Classification and Mapping of Vegetation*, 16.

90. E.g., Frederic E. Clements and Victor E. Shelford, *Bio-Ecology* (London: Chapman & Hall, 1939).

91. Raymond F. Dasmann, "Towards a System for Classifying Natural Regions of the World and Their Representation by National Parks and Reserves," *Biological Conservation* 4, no. 4 (1972): 251.

92. UNESCO, *International Classification and Mapping of Vegetation*, 29.

93. Raymond F. Dasmann, *Environmental Conservation* (New York: Wiley, 1965), 292, 223ff.

94. Dasmann, *Environmental Conservation*, 433.

95. Dasmann and Jarrell, *Raymond F. Dasmann: A Life in Conservation Biology*, 47.

96. In fact, several UNESCO advisors also found fault with IBP's omitting to integrate experts and interests of developing countries in their projects. Chloé Maurel, "L'UNESCO, un Pionnier De l'Ecologie?," *Monde(s)* 1, no. 3 (2013): 185.

97. E.g., see David Pepper, Frank Webster, and George Revill, *Environmentalism: Critical Concepts* (London: Routledge, 2003), 237.

98. Raymond F. Dasmann, *The Conservation Alternative* (New York: Wiley, 1975), 154, 48.

99. "Inter-Agency Consultation Meeting on the Long-Term Programme Based on the Outcome of the Biosphere Conference, 23 and 24 March 1969," RSA SCIBP: NHM, Box 72; also see Holdgate, *The Green Web*, 98.

100. Golley, *A History of the Ecosystem Concept in Ecology*. The MAB consisted of fourteen major themes, summed up in this description: "Man's interaction with terrestrial, freshwater and coastal ecosystems, from polar to tropical zones excluding oceanic systems; natural coastal ecosystems, from polar to tropical zones excluding oceanic systems, natural ecosystems, and systems under various stage of manipulation, transformation and degradation; and large urban systems, considered as ecosystems." Francesco Di Castri, Malcolm Hadley, and Jeanne Damlamian, "MAB: The Man and the Biosphere Program as an Evolving System," *Ambio* (1981): 52.

101. Blair, *Big Biology*, 147.

102. IUCN, ed., *Proceedings of the Twelfth Technical Meeting, Banff, Alberta, Canada 1972* (Morges: IUCN, 1972), 18, 114.

103. IUCN, *Proceedings of the Eleventh General Assembly, Banff, Alberta, Canada 1972*, 144. By 1972, the Ecology Commission, the International Commission for National Parks, and the Commission for Environmental Planning all called for the protection of fragile ecosystems, as well as representative samples of the world's natural ecosystems to understand man's impact on ecosystem deterioration.

104. IUCN, *Proceedings of the Eleventh General Assembly, Banff, Alberta, Canada 1972*, 48ff., original emphasis.

105. Between 1971 and 1973, the International Coordination Council established eight expert panels for the program of the MAB. The panel for Project 8 dealt with con-

servation in particular. Its first meeting was held in late September 1973 at IUCN's headquarters in Morges, and half of the attendees were affiliated with IUCN, including several IBP/CT members, as well as their early adversary, Ray Dasmann.

106. UNESCO, *Expert Panel on Project 8*, 44–46; UNESCO, *Task Force on Criteria and Guidelines for the Choice and Establishment of Biosphere Reserves: Organized Jointly by UNESCO and UNEP: Final Report* (Paris: UNESCO, 1974), 3.

107. UNESCO, *Task Force on Criteria and Guidelines for the Choice and Establishment of Biosphere Reserves*.

108. In 1972, the check-sheet survey had brought in data from about two thousand check sheets that IUCN could operate with. IUCN, *Proceedings of the Eleventh General Assembly, Banff Alberta, Canada 1972*, 51–52.

109. Clapham, *The IBP Survey of Conservation Sites*, 116.

110. Ibid., 115.

111. UNESCO, *Task Force on Criteria and Guidelines for the Choice and Establishment of Biosphere Reserves*.

112. Compare UNESCO, *A Review of the Natural Resources of the African Continent* (Paris: UNESCO, 1963); *International Classification and Mapping of Vegetation*, 20ff. In the meantime, ecosystems thinking also had gathered momentum outside IBP. The expert panel's report of 1973 not only mentions the work of Nicholson's intellectual role model, Elton; it also explicitly refers to the Odums's work on nutrition cycles (UNESCO, *Expert Panel on Project 8*, 11).

113. Clapham, *The IBP Survey of Conservation Sites*; Dieter Mueller-Dombois, *Classification and Mapping of Plant Communities: A Review with Emphasis on Tropical Vegetation* (Chichester: John Wiley and Sons, 1984); UNESCO, *Task Force on Criteria and Guidelines for the Choice and Establishment of Biosphere Reserves*, 50; UNESCO, *Biosphere Reserve Nomination Form*, ed. Division of Ecological and Earth Sciences (Paris: UNESCO, 2013); ibid.

114. Miklos Udvardy, *A Classification of the Biogeographical Provinces of the World* (Morges: IUCN, 1975), 5, 13–15.

115. The MAB consisted of several themed subprojects relating to "man's interaction with terrestrial, freshwater and coastal ecosystems, from polar to tropical zones." In particular, it was stressed that "natural ecosystems, and systems under various stages of manipulation, transformation and degradation; and large urban systems, considered as ecosystems" would be included. Di Castri, Hadley, and Damlamian, "MAB: The Man and the Biosphere Program as an Evolving System," 52.

116. In 1964, the year of the launch of the IBP, the United Nations had established the United Nations Conference on Trade and Development as a permanent intergovernmental body to deal with economic development issues, including scientific research and technological support.

117. Di Castri, Hadley, and Damlamian, "MAB: The Man and the Biosphere Program as an Evolving System," 54; Amadou-Mahtar M'Bow, "Man and the Biosphere," *The UNESCO Courier* 34, no. 4 (1981): 5; Singh, *United Nations Educational, Scientific, and Cultural Organization (UNESCO): Creating Norms for a Complex World*; Worthington, *The Ecological Century*.

118. IUCN, *Proceedings of the Eleventh General Assembly, Banff, Alberta, Canada 1972*, 51–52.

119. Today, the MAB's World Network of Biosphere Reserves continues with 631 nature reserves in 119 countries, representing the world's "major biogeographical regions." UNESCO, *Biosphere Reserves: The Seville Strategy and the Statutory Framework of the World Network* (Paris: UNESCO, 1996); *Biosphere Reserve Nomination Form*; "Man and the Biosphere Programme," accessed 24 March 2019, http://www.unesco.org/new/en/natural-sciences/environment/ecological-sciences/man-and-biosphere-programme/.

 CHAPTER 3

Expertise and Diplomacy

Systems Politics at the UN Stockholm Conference, 1972

Introduction

The United Nations (UN) Conference on the Human Environment held in Stockholm between 4 and 16 June 1972 has entered the history books as the moment when the world community turned the global environment into a topic for international politics and diplomacy. Historians agree that the conference brought about significant changes in how international organizations, national governments, and parts of the public perceived environmental problems and potential ways to solve them.[1] The Declaration of the United Nations Conference on the Human Environment, signed in Stockholm, was to bind UN member states to common environmental goals. Furthermore, the conference laid out an Action Plan for the Environment for the UN General Assembly, recommending activities in environmental assessment, including monitoring and supporting measures.[2] Parts of both the Declaration and the Action Plan remain foundational sources for present-day international environmental decision-making.[3] The conference also resulted in the establishment of the United Nations Environmental Program (UNEP), the first UN organization with its headquarters in the global South.[4] Since its establishment, UNEP, together with the Environmental Fund, which it controls, has become one of the most powerful catalysts of international environmental politics.[5] Moreover, the conference contributed to transnational legal mechanisms aimed at protecting parts of the environment, such as the World Heritage Convention, the Convention on International Trade in Endangered Species of Wild Fauna and Flora (CITES), and the Ramsar Convention on Wetlands, which have been celebrated by the nature conservation community.[6]

Around 1970, when conservationists at the International Union for Conservation for Nature and Natural Resources (IUCN) first heard about the UN's plans for a conference on the protection of the environment, these plans seemed to promise a suitable opportunity to give their work a more visible role in intergovernmental politics.[7] Yet, the UN conference would attract many different types of actors with diverse ideas on what was wrong with the environment. In 1971, with the preparations of the conference in full swing, IUCN President Harold Jefferson Coolidge, an American zoologist, received a letter from his friend Roland Clement, vice-president of the Audubon Society, addressing the growing variety of approaches to environmental protection and the role of conservation therein: "Several different philosophies of conservation are in the air at present, so it is important for us to coordinate especially in view of the fact that the Stockholm Conference may freeze a pattern of approach for some years to come."[8] Clement rightly anticipated that the event would be crucial for the scope of nature conservation and for the influence the members of IUCN would have within the larger domain of environmental problem-solving.

The conference had an important role in defining the scientific approaches and expert roles about which Coolidge and Clement corresponded in 1971. This chapter looks at how environmental problems were formulated at the UN level, how roles were assigned to scientific experts during the years around the Stockholm Conference, and what types of functions IUCN affiliated biologists were able to fill within the institutional framework at and after the conference. Conservationists linked to IUCN, in the wake of the International Biological Program (IBP) and with a new shared scientific approach to ecosystem conservation, tried to position themselves as scientific advisors and experts in the new arena of international environmental politics. On this new stage, biologists linked to IUCN not only had to relate to scientists and theorists from other disciplines and other organizations, but also had to respond to the concerns of national delegates and their supporters. The chapter argues that regardless of large ambitions and seemingly favorable preconditions, the decisions made on environmental expertise during the conference pushed international conservation experts at IUCN into a more confined and isolated position.

Most historical accounts of the conference have looked at Stockholm as the place where country representatives from all regions and political orientations first came together to negotiate definitions, organizational structures, and first regulations regarding the global environment. In particular, the conference has been described in the light of Cold War politics and against the background of growing tensions between the global North and South in environmental and developmental matters. While discrepancies between East and West were felt mainly during the preparation of the conference, authors have shown how during the congress itself, demands for international environmental laws by Northern environmentalists clashed with demands for economic independence and compensation by representatives of newly independent countries from the global South.[9] With respect to this complex political climate in which the Stockholm Conference took place, most accounts have provided a binary reading of the interests brought to Stockholm. These interpretations distinguish between, on the one hand, participants concerned with environmental protection, and, on the other hand, those focused on economic interests in the development of natural resources.[10]

While these accounts aptly discuss the conference's meaning for international diplomacy in a time of political conflict, the small role assigned to IUCN conservationists in the UN's environmental considerations had little to do with the struggles between those scientists and politicians in favor of economic development and those against it. Rather, it was the insistence of IUCN members on a single set of globally applicable ecological rules to underlie all types of development programs that caused opposition. In the course of the conference, global environmental problems would not be defined as lying in the realm of biology, of ecological feedback loops, or in the sound management of natural resources. Instead, regulating environmental systems became a question of social justice, trade regulations, and a case-specific, local application of science and technology to sustain these. At the same time, the intentions of IUCN conservationists to install their organization as the main scientific advisory body to international environmental policymakers clashed with the UN politics of reestablishing trust in their own specialized agencies and regional commissions.

The Rise of Global Systems Thinking

Before turning to the main focus of this chapter—the significance of the role of science-based conservation advice at the Stockholm Conference—it is necessary to look at some of the broader developments that took place in the years leading up to the international gathering. In many ways, the Stockholm Conference was the UN's answer to processes that were already underway in many scientific and political institutions. By 1970, the ecologists in IUCN, the IBP, and the United Nations Educational, Scientific, and Cultural Organization (UNESCO) were no longer the only professional group concerned with the environment. In fact, a growing number of new projects engaged entirely different networks—for example, the microbiologists and chemists in the Scientific Committee on Problems of the Environment (SCOPE), or the economists in the American think tank Resources for the Future Foundation. A strong interest in aspects of what had come to be called the environment had also taken root within different UN specialized agencies. The programs of several of these agencies—for example, the World Health Organization (WHO), the World Meteorological Organization (WMO), and the Food and Agricultural

Organization (FAO)—showed a growing interest in their projects' impact on the environment.[11]

As an intergovernmental organization, the UN was also responding to trends emerging in many of its member states.[12] Several countries had begun to produce annual reports on the state of the environment.[13] A large number of European countries started their own government initiatives to record and protect the environment, and a broadly oriented European Conservation Year was planned for 1970. Simultaneously, the United States established a Center for the Quality of the Environment and passed a National Environmental Policy Act (NEPA), which was to ensure environmental protection in a national framework. Initiatives likes these were backed by a growing public. On 22 April 1970, environmentalists in American schools and universities celebrated the first Earth Day. By 1969, the antinuclear group Friends of the Earth had been founded. The activist group Greenpeace followed in 1971. The environment had become a matter of constituency, at least in what was called the developed world.

These movements were fueled by a growing body of green literature, manifestos, and public reports. In 1962, Rachel Carson's *Silent Spring*, selling 250 thousand copies in its first four months, had triggered a flood of popular scientific literature on the worsening state of the environment.[14] Starting with Carson's condemnation of the careless use of synthetic pesticides such as DDT, concerns regarding environmental and societal threats were closely interwoven in public minds and intellectual arguments by 1970. As described in chapter 1, this was most visible in the bestselling publications of Barry Commoner's *The Closing Circle* and Paul R. Ehrlich's *The Population Bomb* in 1971.[15] Similar to Carson, both authors discussed the negative side effects of modern production on environmental pollution and resource depletion.[16] At about the same time, the Club of Rome, a scientific think tank concerned with international development, published *The Limits to Growth*, which seemed to support the Malthusian hypotheses of Ehrlich and Commoner. The report drew on a computer simulation by the American cyberneticist Jay Forrester and his systems study group at the Massachusetts Institute of Technology (MIT), which ran different scenarios based on five variables: world population, industrialization, pollution, food production, and resources depletion.[17] *The Limits to Growth* exemplified a new global epistemology. By 1970, the advancement of more sophisticated satellite technology from the space programs in both East and West had radically changed the way planet Earth was perceived.[18] While rocket engineering directed Cold War tensions to outer space, the earthly environment was increasingly perceived as a closed system with limited natural resources—a fragile vessel in the vast uninhabitable universe. With all of humankind depending on the continued existence of this *Spaceship Earth*, political leaders were pressured for international regulations to deal with the environment at large.[19]

The preparations for a UN Conference on the Human Environment ran parallel to these developments. In 1968, the same year in which the Club of Rome was founded, the Swedish government suggested that the Economic and Social Council of the UN (ECOSOC) set up a platform for public and governmental involvement in questions on the human environment.[20] The proposal fell on fertile ground. In 1969, a new resolution was adopted at the UN General Assembly to hold a conference that would concentrate the UN's efforts on protecting the global environment.[21] A Preparatory Committee for the conference, composed of twenty-seven member states, international organizations, and nongovernmental organizations (NGOs) such as IUCN, was established in April 1970. Its objective was to identify the main environmental threats and how to circumvent them. The Preparatory Committee, consisting of experts from an array of different disciplines, was responsible for the conference program and the original agenda. In 1970, a small Conference Secretariat was created.

The members of the secretariat were aware that if they wanted to lead a joint effort of country representatives, they had to navigate the diverse opinions about what was wrong with the planet and the ways it was being inhabited. A growing number of publications by specialists who felt both competent and responsible to address environmental problems—natural scientists, engineers, sociologists, economists, to name only a few—provided many, and at times conflicting, interpretations of how the global environmental system could be studied and improved. The disagreement between Ehrlich and Commoner on whether the main problem lay in population growth or the capitalist exploitation of resources was just one example. Growing controversies about the finiteness of natural resources and the power of computer calculations as promoted by the Club of Rome and others presented a second.[22] Therefore, in 1971, the Conference Secretariat called upon a large number of scientists, environmentalists, and public intellectuals to produce a broad and inclusive report on the global state of the environment. The final report, meaningfully called *Only One Earth*, was edited by the British political economist Barbara Ward and the French-born American microbiologist and environmentalist René Dubos.[23] Published a few months before the Stockholm Conference, the report's cover image and subtitle, *The Care and Maintenance of a Small Planet*, underwrote the idea of planet Earth as one vulnerable environmental system drifting through space (figure 3.1). Intended to present in adequate form the diverse social, economic, and political dimensions of the global environment, the independent study served as an important intellectual background document sent to all delegations and representatives.

Figure 3.1. Cover image of the report *Only One Earth*, 1972. Reproduced from Barbara Ward, and René J. Dubos, *Only One Earth: The Care and Maintenance of a Small Planet* (New York: W. W. Norton, 1972), with kind permission of W. W. Norton.

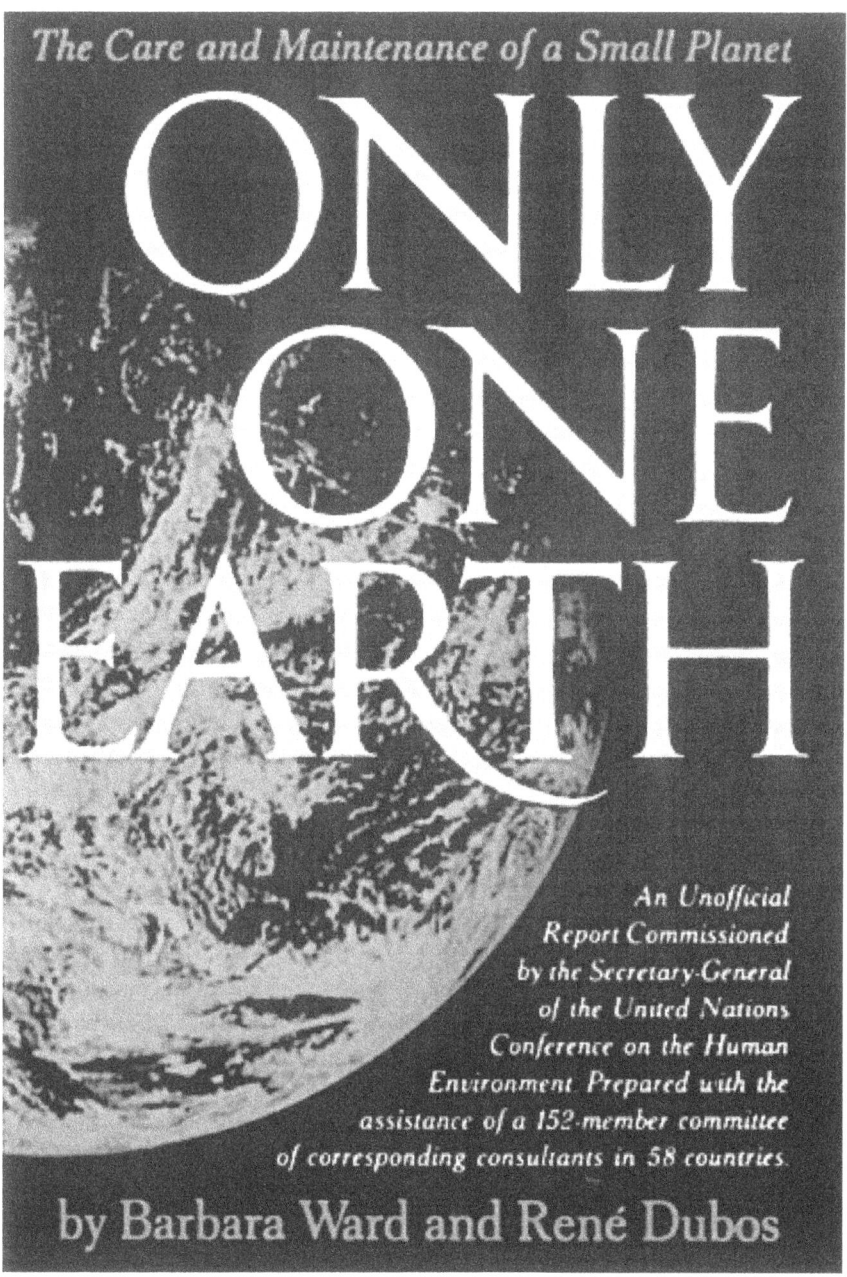

Due to its many and diverse contributors, the study was presented as fair and balanced.[24] At the same time, the texts—condensed, with their original authorship obliterated—were heterogeneous and in many parts contradictory. It is worth disentangling the different interpretations of environmental problems and solutions put forward in the report, as they present a most adequate cross-section of the ideas and opinions hovering over the participants at the conference. Although typifying extreme positions that were often partly blended in the collective study, one can discern at least three different logics within the intellectual framework in which conservationists at IUCN, as well as other expert groups, could position their claims at Stockholm. All three lines of argumentation looked at the global environment as one closed system, and all three focused on human population numbers, potential limits to natural resources, and environmental degradation. Yet each pointed to different sets of causes of, responsibilities for, and potential solutions to the environmental crisis.

The first line of argumentation was mainly concerned with the biological functioning of the environment. In this interpretation, largely developed within the life sciences, the *biosphere* had become an important conceptualization of the global environment. As discussed in chapter 1, the idea of the biosphere, the shared living space of everything and everyone alive, gained popularity in the late 1960s and early 1970s, after UNESCO scientists had organized an international Biosphere Conference, and after the popular science journal *Scientific American*, under the direction of ecologist Evelyn Hutchinson, had published a special issue on the theme.[25] From the three interpretations of the global environment discussed here, the one that centered on the idea of the biosphere was the closest to the reasoning of conservationists at IUCN. That the biosphere featured quite dominantly in *Only One Earth* is further not surprising given the large number of conservation-minded scientists involved in the report, including, for example, the British IUCN member and IBP convener Max Nicholson and his Russian colleague Victor Kovda, by now working for UNESCO.[26] The biosphere was presented in the report as a closed environmental system with limited resources, on which humanity depended.[27] In this conceptualization, the neglect of ecological laws and limits to natural resources posed the most serious threat to the health of the global environment. At the same time, understanding these laws could also help create solutions.

Securing the balance of the biosphere, nature conservation was seen as an important part of environmental solutions. Through the study of balances and cycles, and the preservation of samples and seeds, conservation could secure a "fallback" position for man in the case of natural disasters. In order to study how far the biosphere could be used or modified, some parts of the natural environment needed to be protected as natural laboratories to augment the partial knowledge of the "interdependences of living things and underlying balances of natural order," a passage in the report section titled "Man's Use and Abuse of the Land."[28] While it was thought possible to improve the human condition by working with, instead of against, natural ecosystems, at the end of the day, human beings were believed to be dependent on the only viable planet in space.[29] There were "limits to the amount of manipulation that man can exert upon natural balances without causing a breakdown in the system," another section of the *Only One Earth* report, titled "Strategies for Survival," suggested.[30] Although the original authors remain unknown, the section clearly mirrors the rhetoric of many leading IUCN and UNESCO conservationists at the time, including Coolidge, Gerardo Budowski, Raymond Dasmann, and Duncan Poore.

In many ways, this position corresponded to the Malthusianism of Ehrlich's *The Population Bomb* and the predictions made by the Club of Rome, represented in the committee of consultants to the report by its president, Aurelio Peccei, and MIT systems engineer Carrol Wilson. At Stockholm, IUCN conservationists promoted especially this first conceptualization of the global environment as a closed biosphere. Yet not all consultants put such a strong emphasis on the ecological boundaries to planetary resource management.[31] There are two other systems logics to be found in *Only One Earth*, one inspired by engineering and one based on economic theory. These latter two logics are important because they would come to dominate the discussions at Stockholm. It is to these interpretations that IUCN elites had to adapt their arguments if they wanted to contribute to the environmental discussions at the conference.

The second logic stressed technological solutions, meant to improve the ways societies dealt with their natural surroundings. Whereas natural limits presented an initial problem to the management and use of natural resources, this logic postulated that the human innovative spirit would eventually allow the transcendence of a dependence on the planets' resources. This idea corresponded to the techno-optimist and cybernetic approaches to the environment that proclaimed the dissolution of all boundaries between the natural and the technical, merging knowledge from ecology and engineering. Several passages from *Only One Earth*, especially those concerned with resource and land management, stressed that the natural and technical environment constituted a "single unified system from one end of the cosmos to the other." "In the last analysis," the section called "A Delicate Balance" stressed, "everything is energy."[32] This focus on processes in which different entities with similar functions were quasi-interchangeable had consequences for thinking about environmental responsibilities and the value of the natural world. In this interpretation, engineering and design became important sources of expertise in solving environmental problems. In line with this approach, the extensive section called "Man's Use and Abuse of the Land" discussed technological and engineering solutions to pollution problems, and ideas on recycling used materials or returning them to the ground.[33]

An influential source of inspiration for these ideas was the prominent North American futurist thinker and architect Buckminster Fuller, who brought space cabin–inspired engineering into his design studies.[34] Between 1964 and 1968, Fuller had been developing ideas for a booklet with the telling title *Operation Manual for Spaceship Earth*.[35] In Fuller's version, problems in natural systems could be solved if efforts and labor were focused on engineering technical solutions. This way of thinking contrasted with the biology-oriented biosphere approach that was less anthropocentric and more critical of technological solutions to biological problems. The idea of the human environment as a space cabin that could be controlled and improved at human will was in fact a reaction to the Malthusian pessimism that accompanied some of the biological interpretations that predicted an inescapable ecological disaster. Clever engineering seemingly gave humankind the key to tamper with their own life-support system, in this way designing an optimal living space.[36]

Whereas these two first lines of argumentation focused on the material— the biological or technical—dimension underlying environmental degradation and potential societal consequences, a third logic saw the leverage point for solutions in the economy of natural resources. In this third approach, environmental problems were seen as connected not so much to the availability of natural resources, but rather to their distribution.[37] In 1966, Ward, one of the editors of *Only One Earth*, had published a book called *Spaceship Earth* that emphasized the interconnectedness and interdependencies in global resource supply.[38] The focus in her work and in several passages of *Only One Earth* was not on systems of energy flows, but on problems of unequal access to food, energy and water, natural minerals and metals, and plants and animals.[39] *Only One Earth* was written in a time of economic globalization. A growing number of development experts decried that recently decolonized countries were now confronted with new dependencies on the global market, which ran counter to their goals for political and economic independence.[40] In this respect, questions on natural resources and industrial pollution on a global scale came to be linked to the topics of economic globalization and the need to reform developmental aid policies. The authors of *Only One Earth* had therefore consulted a number of socially engaged economists, including Raúl Prebisch, who had functioned as the secretary general of the UN Conference on Trade and Development (UNCTAD) in 1964.[41] In the late 1940s, Prebisch had become famous in circles of development theorists with the idea that resources always flow from a periphery of poor and underdeveloped states to a core of wealthy and better-developed states, enriching the latter at the expense of the former.[42] By the mid-1960s, Prebisch's criticism on traditional development patterns of technological transfer had become more widespread.

In a similar manner, *Only One Earth* stressed the need not only for political and economic reconsiderations in global environmental, resource, and development questions, but also for a rethinking of the ways in which science and technology were part of both global problems and solutions. This third interpretation shared with the second systems logic a strong confidence in the human capacity to overcome ecological limits. An inspiration for this point might have come from the works of the second editor of *Only One Earth*, Dubos, who in the past had repeatedly pointed to the adaptability of human society.[43] Yet, more than the other two logics, the third interpretation seemed to question the current ways in which expert advice was administered and implemented within global institutions dealing with trade and economic development. It stressed the need of all aid organizations and governments around the world to "adapt to their own local conditions" all forms of advice and aid, given or received.[44] In this way, the third logic was opposed to the common practice of transferring Western science and technology to developing countries without taking into account local demands, traditions, or forms of knowledge. This socioeconomic perspective on the global environment would become especially relevant in guiding decisions at the conference on how environmental expertise and conservation advice was to be managed within the UN system.

All three ways of reasoning through global environmental problems and potential solutions were well represented in 1972 at the UN Conference on the Human Environment. The environment clearly was conceived of as one closed global space, but it remained unclear whether biological or engineering rules applied, or whether the system for global governance itself needed to be overhauled. If IUCN elites wanted to have a voice in the discussions on international environmental policymaking at the conference and after, they had to address these different standpoints. Often this meant presenting their arguments on the global biosphere in a way engineers and economists could relate to.

The "Ecological Rules of the Game"

Leading IUCN conservationists watched the conference preparations with much interest. In many ways, the events at Stockholm seemed to reconfirm the work they had been doing, and conservationists were eager to get involved. Budowski, especially, wanted to make sure that conservation as defined by IUCN—ecologically grounded and involving the management of natural resources—was recognized as an authoritative global voice in the policy discussions at the Stockholm Conference.[45] Yet for IUCN conservationists to present themselves as environmental experts on this new political stage, much skill in diplomacy and argumentation would be required. Coolidge and others had to appeal to many different constituencies, bureaucrats, state actors, and local perspectives, while at the same time promoting a clear expert profile. In parallel, nature conservationists at IUCN faced competition from diverging disciplinary perspectives. Increasingly, claims for expertise on the human environment also came from the social and economic sciences; disciplines such as ekistics—the study of human settlements—did not see ecology, but the management of societies, as central to the solving of environmental problems.[46] While IUCN's environmental diplomacy thus engaged statespersons as well as different types of scientific organizations, IUCN's own scientific philosophy was challenged by new forms of strident environmentalism.[47] As one IUCN member later recalled, newly emerging green activist groups of the late 1960s and early 1970s considered IUCN as part of the "establishment" and insufficiently radical.[48] In turn, IUCN President Harold Coolidge was concerned about IUCN's status as a respectable scientific organization and was eager to set his organization apart from the more vocal activist groups and NGOs who had announced their presence at Stockholm.[49]

From the beginning, IUCN conservationists were keen to collaborate with the official Preparation Committee. Already in 1969, IUCN conservationists exchanged their suggestions for a 1972 conference on "Problems of [the] Human Environment."[50] In 1970, the conference's secretary general Maurice Strong paid a visit to the IUCN headquarters in Morges, Switzerland, as part of his preconference campaigning. According to IUCN's director general Budowski, the meeting resulted in the "mutual satisfaction of excellent understanding" with Strong. Budowski was optimistic that "fruitful co-operation [would] be established in the near future."[51] The key to such fruitful cooperation, many IUCN members thought, was found in their expertise in ecosystem ecology. In the previous years, the Union had increased its emphasis on ecological work, a focus reinforced by the experiences brought back to IUCN by many leading members who had been involved in the Section for the Conservation of Terrestrial Communities of the IBP (IBP/CT). The IBP had understood nature conservation as a discipline for efficient and scientifically planned land use and resource management based on ecosystem ecology. Despite significant discrepancies within IUCN on the importance of this broad definition of conservation, a strong compromise had eventually been found in the framework of ecosystem ecology with a focus on different biogeographical regions. As one member later recalled, by 1971, IUCN had become "by far better placed to contribute to the Stockholm process" than it had been before the involvement of many of its members in the IBP.[52] The new conservation paradigm helped IUCN elites to position themselves as experts of the global environment and to link their knowledge of conservation to the contested topics of growth and development.

This vision of leading IUCN members for science-based conservation, as well as ecosystem research and management, was reflected in the composition of the team of IUCN conservationists who led the negotiations with Strong. IUCN's president Coolidge, by this time already in his late sixties, was a renowned and well-connected scientist who had held positions in the zoology department at Harvard and had directed the Pacific Science Board of the National Academy of Science until 1970.[53] He was joined by his German-Venezuelan colleague Budowski, who promoted a form of conservation strongly rooted in ecosystem ecology.[54] Budowski was now in his forties and had been one of those members of IUCN who, in the late 1960s, had joined Dasmann in lobbying for a closer alignment with UNESCO's plans for the MAB. Dasmann, who had been important for IUCN's cooperation with UNESCO, remained senior ecologist. Next to Coolidge and Budowski, many IBP members, such as convener Max Nicholson of the IBP/CT, IBP's scientific director Barton Worthington, and Nicholson's disciple Poore, were still very active within IUCN. Together, they now promoted a science-based approach to conservation that was linked to ecosystem ecology on a global scale and that respected local particularities. Moreover, they shared a common ambition: they aimed at making IUCN the main authority in scientific conservation and a major player in international environmental policy questions.[55]

This group of conservationists saw IUCN as well suited to advise the UN in environmental matters. At the time, the UN treated environmental problems as part of several other issues related to the management of natural resources, which were dealt with across the different specialized agencies. These included projects on desertification, agriculture and supply, forestry, fishery water supply, and tropical ecosystem management. During the action phase of the IBP, IUCN's conservation principles—discussed, for example, in 1969 at their Tenth General Assembly—also had become increasingly concerned with conservation not only as the protection of wild nature, but as the management of natural resources more generally, so as "to provide for the best sustained social and economic return."[56] By 1970, IUCN's program addressed broad scientific and sociopolitical issues such as world population growth, environmental pollution, environmental planning, and land management as part of a "sound conservation policy."[57] For IUCN, the Stockholm Conference presented a new platform from which its members could apply their ecological knowledge to environmental problems at large, including all underlying "sociological, economic, physical and biological and social factors."[58]

Not only were conservationists trying to apply their ecological expertise to themes relevant to society, they also attempted to engage with other types of experts, especially economists and development specialists. To strengthen this new profile, Budowski was eager to "change the balance of [IUCN] from well-meaning European lovers of wildlife to hard-headed third-world oriented ecologists and economists," which he announced in 1971 in the *New Scientist and Science Journal*.[59] Already in 1968, members of the Conservation Foundation, for instance, had organized a scientific conference at Arlie House, Virginia.[60] The interdisciplinary meeting resulted in a voluminous report, *The Careless Technology*, written with academic scrutiny and building on numerous case studies on the use of "heedlessly" introduced development strategies and technologies that disregarded local ecological particularities.[61] Following this initiative, in 1970, Nicholson, Budowski, former IUCN staff ecologist Lee Talbot, and Peter Scott, an early IUCN member now working for the World Wildlife Fund (WWF), had tried to set up a Working Group on Ecology and International Development. On 24 and 25 September 1970, IUCN and the North American Conservation Foundation convened, under Nicholson's direction, a first such meeting at FAO's headquarters in Rome.[62]

Nicholson's working group had invited development specialists from the International Bank for Reconstruction and Development (IBRD) of the World Bank and the United Nations Development Program (UNDP). The objective of the working group meeting was to assemble views on the content of a proposed guidebook for economic planners, *Ecological Principles for Economic Development*, which Dasmann, together with John Milton and Peter Freeman of the Conservation Foundation, was to draft.[63] Based on the model of the Airlie House Conference report, the guidebook was to contain a number of case studies on successful ecological planning and development from different world regions.[64] The book was written from "the point of view of the ecologist," the preface proclaimed.[65] It was directed, however, at nonacademics and practitioners concerned with development on the national or international level. This was exactly the approach that IUCN conservationists brought to the Stockholm Conference.

In the years leading up to the conference, IUCN conservationists like Budowski, Coolidge, and Nicholson used this broad and societally oriented conception of nature conservation to play an advisory role in the UN system. By now the previously estranged conservationists Dasmann and Nicholson worked closely together, and Nicholson's comments on Dasmann's draft were much appreciated.[66] The final version of *Ecological Principles for Economic Development*, published in 1973, emphasizes ecosystems subject to development pressure, and references the work of both the IBP and the Man and the Biosphere Program (MAB). The content of the book goes far beyond the traditional conservation of natural areas concerned with protecting rare or threatened landscapes or species.[67] One of the most important tasks of the guidebook was to show how various factors playing into development considerations could be compared. In this, environmental and economic factors were supposed to be of equal importance, while the most effective development projects aimed at long-term results and started from an a priori consideration of ecological factors. The central argument to both the Arlie House Conference report *The Careless Technology* and Dasmann's *Ecological Principles* is that development efforts that ignore what Dasmann, Milton, and Freeman had termed the "ecological rules of the game" are bound to suffer adverse consequences.[68] These rules were composed of general conservation principles adaptable to different biogeographical regions, designed to steer all countries' usage and development of natural resources.[69] During the preparations of the Stockholm Conference, IUCN conservationists built on this line of argumentation when promoting themselves as environmental advisors to the UN and its member states.

As explained before, conservationists were not alone in their promotion of biological rules to underlie the successful development of natural resources and economies. Biospheric thinking was widely spread, and, by 1972, the Club of Rome and the MIT Systems Dynamics Group had published their study, *The Limits to Growth*.[70] Many IUCN conservationists supported the systems discourses on the fragility of the biosphere and the need to limit the exploitation of natural resources, stressing the need to strive for a "healthy 'no-growth' world."[71] Similarly, the utilitarian, growth-critical ideas of ecologist Garret Hardin, the author of the *Tragedy of the Commons*, found support in IUCN circles.[72] Conservation elites at IUCN shared with Hardin and the Club of Rome members a critical attitude toward purely technical solutions to environmental problems,[73] and the "naïve faith in the ability of science and technology to develop new resources, and even new environments."[74] IUCN's president and director general instead strove for new expert alliances with enlightened economists such as Ward. Together, so they hoped, they could use the existing intergovernmental structures of the UN to convince political decision-makers and heads of states of the necessity to work with, not against, the ecological processes of the biosphere.

Previously, IUCN had had only limited success in attracting the attention of political organizations and governments.[75] Around 1970, the UN's new interest in the environment seemed to present a promising opportunity to make up for previous neglect. In 1969, when the plans for the Stockholm Conference were first discussed at IUCN's Tenth General Assembly, Budowski had praised the "rising tide of conservation interest on the part of governments and technical assistance organizations" in the development sector, such as the UNDP, UNESCO, and FAO.[76] These UN agencies seemed key to granting IUCN elites access to expert positions within the UN and to earn them the recognition and respect of governmental actors and state delegates.

With existing links to UNESCO, IUCN conservationists turned to FAO and ECOSOC, responsible for UNCTAD and the World Food Program, respectively, for further cooperation. Both agencies presented international questions on the development of natural resources as part of their competences, and were thus perceived as potential allies. Already in 1970, Budowski had approached FAO, hoping "that IUCN, as a non-governmental scientific organization with an easy—and . . . quick—'scientist to scientist'" approach, could contribute a great deal to what he saw as a major objective of the planned UN conference: the introduction of much more basic and applied ecology and "conservation components into the major development schemes."[77] In January 1971, Coolidge and IUCN Deputy Director General Frank Nicholls additionally made contact with Philippe De Seynes and Robert Muller of the UN Office of the Secretary General; Joseph Barnea, director of the Resources and Transport Division at UN; and Guy Gresford of the Science and Technology Department of ECOSOC, asking for a larger involvement of IUCN scientists in ECOSOC's work on resource-development politics in the months before the conference.[78]

Networking helped to strengthen both IUCN's unofficial and official position during the preparation phase of the conference.[79] Behind the scenes, IUCN and other expert groups began to form coalitions, in this way hoping to avoid having to go through the higher, highly bureaucratic sections of the UN organization.[80] Former IBP contacts proved to be useful in strengthening the connections with national delegations and diplomats. During the first session of the Preparatory Committee, for instance, Victor Kovda, now representing the USSR, stressed that IUCN was an "organization that could give valuable contribution[s] on the preparations" for the Stockholm Conference.[81] In late 1970, it was agreed within the Preparatory Committee that IUCN, as one of the few scientific NGOs invited to the conference, would work closely with the Conference Secretariat.[82]

Intending to make the most of this opportunity, in 1971 Coolidge chose the environmental law expert Richard N. Gardner as the new IUCN representative to the UN.[83] Gardner was a high-carat addition to IUCN's mission. Educated at Harvard, Yale, and Oxford, he had served as the U.S. delegate to sessions of the FAO and WHO, as the vice chairman of the U.S. delegation to UNCTAD, and had acted as John F. Kennedy's deputy assistant secretary of state.[84] Through Gardner, IUCN would become quite influential in the Conference Secretariat. At the same time, the secretariat doubtlessly recognized the benefits of IUCN's involvement. In particular, the secretariat was interested in IUCN's large network and knowledge of ongoing environmental initiatives.[85] In April 1971, Gardner began to draft a paper for the Conference Secretariat dealing with postconference institutional arrangements regarding the environment.[86] This gave IUCN's leadership the chance to design, at least on paper, a central position for IUCN to give the UN advice on environmental monitoring, research, standard setting in public education, and the settlement of transnational environmental disputes. In this blueprint, IUCN, "the world's only international organization of broad competence concerned entirely with problems of the environment," would play a dominant advisory role within the UN's future environmental efforts and decision-making procedures.[87]

Given these first successes and new alliances, it came as a surprise to many when IUCN's seemingly central position within the conference hierarchy began to change in the months prior to the event itself. In particular, it proved increasingly difficult to interest other UN agencies in the ecological advice that IUCN hoped to provide, and closer institutional links with FAO and ECOSOC never materialized.[88] In 1972, many important decisions would be made over Coolidge's, Budowski's, and Gardner's heads. In this, IUCN was not alone. At the conference, many other NGOs were disappointed to find that the conference was structured in a way that limited their participation to a minimum.[89] At Stockholm, most conference time was dedicated to the committee meetings, where all governments were present, taking time away from the plenary sessions, which gave some more room to NGO participation.[90]

In general, the conference organization, although initially interested in a broad spectrum of opinions, kept the official diplomatic conference and events for NGOs largely separate (figure 3.2). NGO activities were convened as a sep-

Figure 3.2. A general view of the opening meeting of the conference at the Folkets Hus, 5 June 1972 © UN Photo/Yutaka Nagata. The Folkets Hus had been prepared to accommodate the large number of UN representatives attending the different sessions of the conference.

arate event at an external location: the Environmental Forum. According to the official information notes, the Forum was there to "stimulate continued debate and a deepened engagement in environmental questions."[91] The Forum indeed became an important platform for many NGOs. It received over 180 proposals for the generous exhibition space of a thousand square meters. Yet the separate organization also turned it into an event that was quite different from the conference (figure 3.3). For the conference organizers, the plans for the Forum most certainly also served as an appeasement instrument to keep radical activist groups out of the conference rooms.[92] At the Forum, IUCN took on a dominant position compared to other NGOs, such as the WWF or Friends of the Earth.[93]

The work that IUCN had done prior to the conference, based on their conceptual work on ecological guidelines for development experts, found little attention at the conference itself. At the plenary sessions, NGOs under the leadership of IUCN clearly used a rhetoric of limited resources, vulnerable life-support systems, risk, and the restriction of the need for an "economic system not exceeding the renewable resources and the carrying capacity of the environment."[94] Yet, during the two-week conference, many suggestions that NGOs, including IUCN, had brought to the conference preparations were overturned. A draft of the UN's Declaration for the Environment, intended to

Figure 3.3. The Environmental Forum at the National School of Art in Stockholm, 7 June 1972 © UN Photo/Yutaka Nagata. Plenaries at the Forum were crammed into auditoriums.

"inspire and guide the peoples of the world in the preservation and enhancement of the human environment," which had been formulated with the help of the NGOs present at the conference, was turned down by representatives from the People's Republic of China.[95] In the process of formulating a new document, four new principles entered the final Declaration. Importantly, two of these principles addressed that environmental measures needed to recognize the economic interests of developing countries.[96] As Indira Gandhi, India's Prime Minister, famously defended, poverty was the biggest polluter and needed to be eradicated before other environmental concerns could be targeted effectively.[97]

Conservationists at IUCN had originally believed that their cooperation with UNESCO, ECOSOC, and FAO would reward them, as scientists, with the access to national delegates and diplomats that they needed to address ecological development problems effectively on the ground. They did not see, however, that the UN system at the time was itself under severe criticism from within the development sector. Many member states were discontent with the status quo of international development politics, based on ideas of modernization through the dissemination of science from the global North to other geographical regions, without adapting these approaches to local circumstances.[98] In 1969, this type of criticism of the UN's development politics had been central to a report by the World Bank's Pearson Commission on International Development, called "Partners in Development," which moved the international focus toward humanitarian issues, promoting the growing integration of social and economic demands of the South into the international agenda.[99]

In the early 1970s, the Pearson Committee's report provided the backdrop to reform plans for the UN's development politics, suggesting a restructuring of development, trade, and the mechanisms for scientific and technological assistance. These calls for reform found much support by intellectuals and political leaders in the developing South itself. Many young independent nations, collectively forming the so-called Group 77 of developing nations, supported the Pearson report's message.[100] During the discussions at the Stockholm Conference, representatives from countries such as Brazil, India, and China firmly linked questions on the global environment to reform plans for the international aid system, trade, and development. Together representatives from these countries insisted on their national jurisdiction in both sectors.[101] Policymakers at the conference increasingly considered the strong international advisory role in global environmental questions demanded by leading IUCN conservationists like Budowski and Coolidge as politically undesirable. Both conservationists' demands for scientific authority and the idea of universal ecological rules for conservation and development seemed incompatible with the changing ideas on scientific and technical assistance within the UN. In fact, UN administrators, led by Strong, were developing their own, opposite ideas on how environmental problems were to be tackled within the intergovernmental diplomatic system of the UN.

The Politics of Environmental Advice

The UN Conference on the Human Environment was a large-scale endeavor attracting many different stakeholders and interest groups. Yet the original organizers of the conference also had ambitious objectives of their own. At Stockholm, crucial differences between the positions of organizers and participants of the conference were obscured by a thick layer of environmental discourse and diplomatic talk. Looking beneath this layer, however, one finds that the UN officials involved in the conference, like Strong, followed plans that at times clearly opposed the idea of universal conservation guidelines for development as defended by the IUCN scientists. In fact, members of the Conference Secretariat, especially Secretary General Maurice Frederic Strong, demanded an interpretation of environmental problems in terms of social reform, distributive management, and technological solutions. Individual governments, Strong believed, could solve environmental problems under the supervision of UN specialized experts. Soon, two related points of conflict emerged between the IUCN elites and the organizers of the UN conference. One concerned the type of expertise that was needed to manage the environmental system; another pertained to the authority of independent scientific NGOs such as IUCN.

In order to understand the parameters according to which expert positions were assigned at the Stockholm Conference, it is helpful to look at Strong and his ambitions in more detail. As the conference's secretary general, Strong was a driving force behind the organization of the conference, as well as the decisions made on policy mechanisms and program points. At the same time, as a high-level UN representative, he advocated a change in attitude toward the international dissemination of development aid and scientific expertise, and a move away from centralized and technocratic cultures of advice by increasing regional cooperation with the UN agencies. Accounts of Strong's persona usually repeat two things: first, that he knew what it meant to be poor, and, second, that he was well traveled and had excellent diplomatic skills.[102] Yet also important in the context of this chapter are his strong interests in internationalism and the links between natural resources and development. His interest in the nonhuman environment came from a very different background than that of Budowksi, Coolidge, or Dasmann. Traveling for different oil companies, the self-made Strong had observed a range of local projects for resource development and extraction firsthand.

Born in 1929 into deprived circumstances in Oak Lake, Manitoba, Strong had left school at age fifteen to become a fur trader and then joined the army.

At seventeen, he began to work for various oil companies, first in Canada, then in other parts of the world. In the meantime, he taught himself bookkeeping and a variety of languages. His oil career and his reading in geology books fueled his interest in the global trade of natural resources. Wanting to pursue a career in commerce, he decided to make up for his lack of formal education by traveling the world, spending a lot of time in Africa. His experiences, paired with his strong Christian beliefs, contributed to his interest in development and internationalism, and his concern about socioeconomic inequalities in a globalizing society.[103]

Early on, Strong had been fascinated by the potential role postwar international institutions, such as the UN, could play as fora in which newly independent countries had a chance to position themselves in the international world order.[104] During the 1960s, Strong made several attempts to attain a position at the UN or one of its agencies, but had been unsuccessful due to unfortunate circumstance and a lack of credentials.[105] Strong instead returned to building up a business career for himself. Successful in the oil and energy business and with international experience, in 1965 Strong was asked by the Canadian prime minister to become the director general of the Canadian International Development Agency (CIDA). This function eventually provided Strong with another chance to launch a UN career when his unusual background and his work for CIDA attracted the attention of UN Secretary General U Thant and ECOSOC's Under Secretary Philippe De Seynes. De Seynes especially found in Strong a diplomatic ally who shared his wish for a peaceful decolonization process and the political and economic integration of the global South. In 1971, the devout Canadian autodidact was promoted to the position of secretary general of the Stockholm Conference on Thant's wish.[106] Strong gladly accepted. In recent years, he, like De Seynes, had followed with some concern the demands by a number of newly independent countries to reform the international aid and trade system. In 1974, when asked by an interviewer of *Delegates World* to look back to the challenges of Stockholm and its aftermath, Strong explained, "My problems have come from within the UN itself. . . . I say this from a pro-UN point of view. . . . This system's in bad trouble. It needs very major overhaul."[107]

To Strong, the Stockholm Conference presented a chance to bring estranged governments together while restoring their trust in the international importance of the UN.[108] At the many UN events where he presented the plans for the Stockholm Conference, his rhetoric always linked in with the goals of the Second Development Decade to improve international economic relations. What was needed, according to Strong, was "reconciliation between the needs for material progress" and "fundamental social goals," between "short term advantage for some and the long term interests of all men and nations."[109] The success of the UN system, he believed, depended on the success of the Stockholm Conference to reconcile those member states divided over environment and development questions.[110] In this, the conference's key concept of the "Human Environment" could help redefine the relations between science and policymaking by providing a management framework for the environment that incorporated local, regional, and national development objectives.[111]

Important, in this respect, is that Strong's interest in development reforms, paired with the philanthropist ideology of a just natural resource management, led him to conceptualize environmental solutions less in ecological, and more in ethical, terms. Strong himself did not oppose the idea that there were certain limits to existing natural resources. He explained that it was the "natural resource base" of water soil, forest, plant, and animal life that provided the "principal basis" for development. Therefore, resources required care, proper use, and careful extraction, Strong later explained in an interview with the Canadian Broadcasting Corporation in 1974.[112] When in 1971 the MIT Systems Dynamic Group, working with the Club of Rome, appeared on the radar of the conference's Preparation Committee, Strong was eager to have its work as background material for the Stockholm Conference.[113] This changed, however, after the publication of *The Limits to Growth*. Although the idea of constraints to the environment and natural resources was not new, what was perceived as controversial about the Club of Rome's study was its presumed patronizing tone toward developing countries. For some, it seemed to suggest that, based on the physical limits of the environment, one set of rules was to govern the fate of all the people on the planet, irrespective of their economic situation. Delegates from developing countries, in particular Brazil and China, reacted with hostility to the report that seemed to forbid growth altogether.[114]

Also UN development experts at ECOSOC were concerned that the concept of a global environment or related discussions on overpopulation and restrictions to economic growth could be interpreted as a form of scientific imperialism.[115] This notion played out as problematic for IUCN conservationists and the inflexible politics that seemed to be implied in their insistence on the ecological rules for environmental manipulation and the universal applicability of their conservation principles. Within the framework of the conference, a single, global, and potentially growth-restricting policy was unthinkable should the interests of the global South be retained. At the first day of the conference, World Bank President Robert McNamara emphasized that development and the betterment of the living conditions in developing countries could not be improved without regional economic sovereignty.[116] This was repeatedly stressed in the rhetoric by UN officials at the conference.

In their attempts to reconcile national agendas and the idea of a closed environmental system, Strong, De Seynes, and others could draw on the criticism that had been voiced in reaction to the Club of Rome report. Although the report was in fact quite nuanced, including scenarios that allowed for a changing technological infrastructure, the bio-deterministic scenarios especially caused hefty counterreactions. A dominant type of critique pointed out that the predications made by the study were based on the assumption that neither the socioeconomic model dominant at the time of its publication nor the technological paradigm would change during the period that it discussed.[117] For Strong, those two factors needed to be transformed. In his opening statement to the conference, Strong explained how "our physical interdependence" on the planet required "action to achieve new dimensions of economic, social, and political interdependence."[118] This statement clearly opposed the fundamentals of IUCN's line of argumentation. The problem for Strong was not the biosphere in its integrity and diversity, but the way that global society was managing its natural resources. Rather than IUCN's ecological approach, which recommended development measures based on the physical limits of ecosystems, Strong promoted an approach that treated society and environment as systems that could be managed though social innovation.[119]

Like Ward and Dubos in *Only One Earth*, Strong could draw on a whole range of systems literature to support his ideas on the socioeconomic management of the environment. Notably, Strong chose the work of the systems thinker and techno-optimist John McHale.[120] In Strong's statements during the conference preparation phase, one can find much of McHale's intellectual influence. McHale, at the time the director for Integrative Studies at the State University of New York and principal organizer of the Future Studies Program, was a keen follower of the American industrial designer and spaceship ecologist Buckminster Fuller, whose ideas had been referenced in several sections of the *Only One Earth* report. McHale combined the technoscientific optimism of his mentor Fuller with calls for social reform and new development politics that matched Strong's ideas. Two of McHale's publications became part of Strong's "Environmental Fact File" that he used to prepare the conference. McHale's *Ecological Context*, published in 1970, discussed a variety of "limits" related to society, the human psyche, and natural resources.[121] More radical even was McHale's *The Future of the Future* from 1969.[122] In *The Future of the Future*, McHale elevated discussions on limits of the natural environment to a higher moral level: the environment could largely be manipulated through science and technology. Yet, although in small scale these manipulations had done little harm, the current magnitude of their use and their consequences required new "value commitments to specifically preferred and possible futures in human terms."[123] In this way of thinking, the main environmental problem was not the inevitable physical limits of the environment. It was the "'software,' or social thinking," in development approaches that was "less than adequate" and in immediate need of "reevaluation and redesign."[124]

This last point is important because it contrasted with the IUCN conservationists' cause in crucial ways. What McHale suggested was not the need for a new unifying scientific approach for the environment in the form of ecosystem ecology as proposed by IUCN elites. Instead, he proposed a new socio-economic ethics that was to guide different perspectives in decision-making processes related to the environment. Under the umbrella of social reform, McHale suggested separate and specialized expert bodies without granting IUCN conservationists or any other scientific expert group the superordinate coordinating role they demanded. On top of this, McHale promoted a form of techno-optimism that allowed for unrestricted economic growth through adapted technological development. Rather than working with the ways of nature, McHale predicted that "at the point, then, where man's efforts reach the scale of potential disruption of the global ecosystem, he invents precisely those conceptual and physical techniques that may enable him to deal with the magnitude of a complex planetary society."[125]

McHale's line of reasoning fed directly into Strong's rhetoric at the conference. It could easily be integrated into proposals for UN reforms regarding the integration of local politics in scientific and technological assistance, which could appease both developed and less developed countries. At the conference itself, in his address to the different government delegates, Strong shifted the focus away from natural science approaches to "*social invention*, to the re-moulding and reshaping of our institutions, organisations and value systems," for the "maintenance of our emerging planetary society."[126] Ecological theory was kept out of the conference rooms. At the same time, McHale's arguments for separate and decentralized expert bodies fit in well with Strong's plans to revitalize the UN agencies.[127] Both emphases explain the limited access granted to IUCN elites at the Stockholm Conference.

Next to these philosophical differences, discrepancies between Strong, Coolidge, and Budowski became most apparent in the discussions on how environmental problems were to be dealt with within the UN's organizational structure from 1972 onward. On 8 July 1971, an expert committee on postconference institutional arrangements met to discuss the different ideas that had been brought forth. A draft by IUCN representative Gardner on the "International Organization and the Human Environment: A Discussion Outline" was reviewed together with one designed by Strong himself. At this meeting, the differences between the two proposed arrangements were apparent to everyone. IUCN's proposal stressed the need for ecological advice by experts in independent scientific NGOs, led by the conservationists at IUCN. In contrast, Strong was eager to show that the existing UN framework could be reformed, so that the UN itself would be capable of dealing with environmental issues. Within the UN system, a firmly integrated unit should oversee environmental policymaking. Yet, rather than working with one set of conservation guidelines, Strong hoped to strengthen the UN's structure and the national support for the UN's specialized agencies by dividing the environmental work according to the different competences already represented within the UN system.[128]

UNEP and a New "Wildlife Bias" for IUCN

The new environmental unit of the UN, UNEP, was founded in 1972 as a direct result of the Stockholm Conference. Strong was appointed as UNEP's first director, and UNEP's program, set down in the Action Plan published in the Stockholm Conference report, was composed in close alignment with Strong's suggestions.[129] UNEP did not constitute a new specialized agency and had much less executive power compared to FAO or UNESCO.[130] Instead of carrying out their own projects, the members of UNEP's governing council were expected to index, classify, and oversee existing projects and activities on aspects of the environment led by the UN member states and specialized agencies.[131] Nevertheless, as a catalyst for environmental projects, UNEP seemed to become an important player in the field of environmental policymaking and diplomacy. With UNEP in place, however, nature conservation advice and IUCN as a scientific, nongovernmental expert organization remained confined to a rather limited role, concerned with traditional conservation topics, such as the protection of endangered wildlife.

As executive director, Strong imposed his philosophy on the composition of UNEP's secretariat. The experts that Strong and the governing council approached in 1974 at the council's second session were neither much interested in ecology nor in conservation-related topics. In fact, few of them had a background in the natural sciences. Strong believed that the experts needed were manager types with a firm training in the social and economic sciences, "generalists with the integrative grasp, broad technical base, systems training and international orientation needed to understand complex trans-disciplinary problems."[132] The main purpose of UNEP, after all, was to "extend the limits of social capacity to cope with environmental problems."[133] It is not surprising, then, that in the staffing of the UNEP's secretariat, Strong mainly drew on business people, politicians, and international relations specialists, the sort of experts with whom he was acquainted from his previous positions at CIDA and Petro-Canada.[134] From the moment Strong became involved with the conference preparations, he had made particular use of the UN development network. In 1971, Ward had established the International Institute for Environmental Affairs (IIEA), which functioned as an independent policy think tank.[135] Soon after the conference, the institute became involved in advising Strong as to which environmental priorities the UN should focus on.

In 1973, the IIEA and the Aspen Institute, under Strong's instruction, formulated a priority research program for UNEP and the Environment Fund, both established at the conference. Similar to McHale's proposals, the Aspen Institute's environmental agenda, as summarized by one of its advisers, was to facilitate a better understanding of the "Issue of Social Choice," and thus the institutional, economic, and technical dimensions of environmental issues, rather than biological limits.[136] McHale himself was one of the main experts consulted.[137] In general, Strong's choice of experts was decisive for the early program of UNEP. Rather than ecological research into biological resource limits, the just distribution of natural resources, their economic management, and ways to prevent industrial pollution came to be seen as the crucial topics for environment research.[138] This approach had major consequences for the further distribution of environmental tasks and advisory positions, including those related to the conservation of nature and natural resources.

Following this Action Plan, the foundation of UNEP was accompanied by the establishment of a voluntary fund, to which especially affluent Northern countries, such as the United States, were expected to contribute significantly in the first years after the conference. By the second half of 1972, the established Environmental Fund had raised US$60 million and planned to raise US$100 million by the end of 1977. For the environmental sector, this was a significant sum, and it amounted to about a hundredfold of IUCN's basic annual budget at the time.[139] Yet, with UNEP's institutional role firmly integrated into the existing intergovernmental structures of the UN, there was not much latitude in spending. The Fund was presented as the "institutional machinery" for governments "to make [environmental] plans a reality" and thus seemed to be firmly coupled with the UN membership status of governmental actors.[140] Additionally, as was clear from the beginning, a big part of the money would go to the UN's "operating entities, including the specialized agencies of the UN system."[141] With IUCN falling into neither category, conservationists were dependent on Strong's goodwill and the organizational role they were granted. Yet, many of the tasks Coolidge and Budowski had hoped IUCN members to fulfill were given to other UN agencies. IUCN's role was defined narrowly and focused on traditional conservation topics they had long tried to transcend.

The Stockholm Action Plan had featured two main objectives that were now integrated into UNEP's program: environmental management and environmental assessment.[142] Neither required IUCN's conservation expertise. Environmental management included topics mostly related to the work of FAO and UNESCO, such as the planning of agricultural progresses, resource and water supplies, and the planning of human settlements. Also in soil, water and forest conservation, and the protection of tourist wildlife sites, leading roles were assigned to members of FAO and UNESCO rather than IUCN.[143] A special emphasis was placed on the development and management of pest control, the productivity of tropical hard woods and fibers, water reprocessing, arid land recovery, damn building and irrigation, early climate research, indigenous building, natural disaster warning systems, and human disease control.[144] Assessment and environmental monitoring had been two of the topics to which IUCN elites had hoped to contribute at the conference. Yet, the type of mechanisms that UNEP wanted to install went far beyond the experience of IUCN's members. Rapid developments in computer and satellite technology, as well as the perceived need for truly specialized scientific bodies, brought with them different types of potential advisors for diverse environment-related tasks. In the years after the conference, UNEP's focus on environmental and development diplomacy would demand a scientific know-how that was scarcely represented within IUCN, such as the conservation of genetic resources. Here, FAO and UNESCO scientists unrelated to IUCN would take the lead.[145]

Within UNEP, then, the nature conservation advice provided by IUCN became defined narrowly as the protection of "the heritage of wildlife and its habitat."[146] This definition of conservation did not take much of the new and broader program of IUCN into account. For some conservationists, including the Union's president, Coolidge, the focus on the protection of wildlife was not completely unacceptable.[147] Leading up to the conference, Coolidge had tried to use his connection to the American Committee for International Wildlife Protection to make IUCN an active participant in the World Heritage Convention.[148] IUCN members moreover managed to present the Ramsar Convention on Wetlands as well as the Convention on International Trade in Endangered Species for signature at the conference. By and large, however, Coolidge's approval for an approach to nature conservation that focused on wildlife and nature reserves did not comply with the overall agenda of IUCN elites at the time of the conference. According to this agenda, nature conservation was to "embrace the entire nature complex," and the new "wildlife bias" imposed on the responsibilities of IUCN by the Stockholm Conference placed "too little emphasis on ecology, forestry, etc.," just as leading representatives of IUCN member organizations had feared.[149]

At IUCN's Eleventh General Assembly of 1972, held in Banff National Park in Alberta, Canada, shortly after the Stockholm Conference, IUCN members who had been part of the IBP/CT section objected that cooperation in the World Heritage Convention, which treated parks as monuments, would compromise IUCN's "scientific work" on national parks, fencing off protected areas rather than making them available for ecosystem research.[150] In general, IUCN's ambitions of the early 1970s went well beyond a limited focus on the protection of wildlife. To the minds of leading conservationists within IUCN, including Budowski and Dasmann, it was mainly their ecological approach to the conservation of natural resources that made them experts on the environment at large. They still hoped that with additional funding by UNEP for the years between 1973 and 1975, a broader program for environmental planning and resource management could be achieved.[151]

The anticipated financial assistance did not materialize, however.[152] Strong had supported IUCN by helping to secure a grant from the Ford Foundation, but only until UNEP was founded. The smaller grant IUCN received from then onward—US$300,000 instead of US$650,000—was needed to finance the organization's extended secretariat alone. IUCN would receive further financial support from UNEP, but the projects funded corresponded to Strong's narrowly defined understanding of nature conservation. UNEP's support was mainly meant for the development of national parks in the global South, as well as surveys of critical marine habitats in the Mediterranean Sea. The amounts involved were limited. Not only did the lion's share of UNEP's financial injections go to FAO and to UNESCO—in total over US$5 million—but this money also funded projects on themes that IUCN conservationists regarded as within their own sphere of expertise, such as ecological research into different types of ecosystems.[153]

Alternative funding channels helped little to strengthen a holistic, research-oriented conservation program. Most basic funding for IUCN was still supplied by the WWF. In 1974, IUCN received US$1 million from the WWF for basic administrative operations. Another US$300,000 was granted to the Species Survival Commission's (SSC's) program, the IUCN commission that, with its focus on endangered species, came closest to the WWF's own idea of conservation. As a consequence, the SSC became a very dominant component of IUCN.[154] In 1977, Coolidge's wife explained in a letter to Talbot that the "SSC practically is IUCN with 900 people involved one way or another." According to her, it had "virtually carried IUCN during the past few years."[155] Naturally, this had only strengthened the focus on the protection of wild and threatened nature in IUCN's program. All in all, and contrary to the IUCN elite's initial expectations, the integrated approach to a broadly defined conservation, including resource management, research, and monitoring, which had become inherent to the organization's self-image, had not allowed IUCN scientists to play an active part in UNEP's projects. Instead, the funding that IUCN received led to a stronger focus on traditional conservation topics, such as the management of national parks and the protection of endangered species in limited geographical regions.

Conclusion

State of the world concepts are formulations of the ways in which people see the world around them. Many of these concepts overlap one another, and some of them are contradictory; all are, in some measure, interrelated. [These] environmental concepts, even when unarticulated or unrecognized, nevertheless underlie all environmental policies, . . . [and] inchoate as they sometimes are, affect our behaviour.[156]

These words by IUCN member Lynton Caldwell, expressed a few weeks after the Stockholm Conference at IUCN's Twelfth Technical Meeting, nicely capture how decisions on environmental expertise in fact depend on particular conceptualizations of what nature is and which environmental problems need to be solved.[157] A similar entanglement of ways of thinking about nature, strategies for environmental problem-solving, and forms of expertise also underlay the roles given to IUCN conservationists in the wake of the Stockholm Conference. At the time of the conference, innovations in computer technologies and engineering contributed to the emergence of diverse systems thinking approaches. These different approaches shared an understanding of the earth as a whole of interlinked and dynamic processes. Yet, we have seen that different types of systems thinking caused discrepancies between IUCN conservationists and the organizers of the conference regarding the distribution of international environmental expertise.

In the course of the 1960s, scientists at IUCN had defined their approach to conservation and their sphere of competences as based on ecosystem ecology, with the emphasis on locally adapted yet broadly valid conservation guidelines, implemented with the help of intergovernmental agencies and institutions. In Stockholm, however, the global environment *problematique* was not defined in terms of biological systems. Rather, shaped by postcolonial politics and plans to reform the international mechanisms of scientific and technical assistance, environmental problems were defined in terms of socioeconomic relations, aid reforms, and a lack of trust in the UN system. For addressing such problems, ecology as promoted by IUCN conservationists was not considered the right type of science. Moreover, after the foundation of UNEP in 1972, environmental NGOs such as IUCN were granted little authority within the UN system. The Stockholm Conference did provide leading IUCN scientists with a venue to negotiate three conservation conventions, but these conventions were based on a rather limited idea of what nature conservation was to entail. Similarly, UNEP mainly funded those IUCN projects that aimed at traditional conservation topics, such as the protection of wildlife and the establishment of nature reserves in lesser developed regions.

Looking at how IUCN conservationists' competences were defined from outside the field of conservation itself, this chapter has generated new knowledge on an important early moment in the distribution of environmental expert roles in the "international decision-making space."[158] In the early 1970s, the IUCN elite's attempt to institutionalize their advice and expertise in the international organizational landscape experienced significant constraints. In the following years, leading IUCN scientists would have to redefine and readjust their own role within the institutional context that the Stockholm Conference had brought about. After the conference, the definition of resource politics in socioeconomic terms, and the understanding of nature conservation advice as related to wild nature, marked a conceptual separation between societal and biological expertise related to the environment. As will be discussed in chapter 4, the need to bridge this division between the social and the natural consequences of environmental degradation significantly shaped conservation strategies pursued by the leading members of IUCN in the years to come.

Notes

1. Frank Uekötter, *Naturschutz im Aufbruch: Eine Geschichte des Naturschutzes in Nordrhein-Westfalen 1945–1980* (Frankfurt am Main: Campus, 2004); Robert V. Bartlett, Priya A. Kurian, and Madhu Malik, *International Organizations and Environmental Policy* (Westport: Greenwood Press, 1995); Stephen Bocking, *Ecologists and Environmental Politics: A History of Contemporary Ecology* (New Haven: Yale University Press, 1997).
2. Stanley Johnson, *UNEP: The First 40 Years; A Narrative* (Nairobi: UNON/Publishing Section Service, 2012), 21.
3. Pierre-Marie Dupuy and Jorge E. Viñuales, *International Environmental Law* (Cambridge: Cambridge University Press, 2015), 9–11.
4. United Nations General Assembly, "2997(XXVII). Institutional and Financial Arrangements for International Environmental Co-operation" (1972), accessed 19 March 2019, https://documents-dds-ny.un.org/doc/RESOLUTION/GEN/NR0/270/27/IMG/NR027027.pdf?OpenElement.
5. Lynton K. Caldwell and Paul S. Weiland, *International Environmental Policy: From the Twentieth to the Twenty-First Century* (Durham, NC: Duke University Press, 1996).
6. Martin W. Holdgate, *The Green Web: A Union for World Conservation* (Gland: IUCN, 1999), 144; Raymond F. Dasmann and Randall Jarrell, *Raymond F. Dasmann: A Life in Conservation Biology* (Bloomington: Xlibris Corporation, 2000), 99; Interview with Lee Talbot, 11 September 2014.
7. IUCN, ed., *Proceedings of the Tenth General Assembly, New Delhi 1969* (Morges: IUCN, 1969), 169; also see Rachelle Adam, *Elephant Treaties: The Colonial Legacy of the Biodiversity Crisis* (Lebanon, NH: University Press of New England, 2014), 86–90.
8. Ronald Clement to Harold Coolidge, 7 May 1971, HUA HJC: HUG (FP) 78.20, Box 8, Folder "IUCN-WWF and Relate International Conservation 1971 IUCN-International Relations: UN Conference on the Human Environment-plans" (2 of 2).
9. Authors have described how, at the last moment, several countries of the Eastern bloc had decided to boycott the conference because the GDR had been denied participation, according to the Vienna formula; e.g., Kai F. Hünemörder, "Environmental Crisis and Soft Politics: Détente and the Global Environment, 1968–1975," in *Environmental Histories of the Cold War*, ed. Daniel McNeill and Corinna R. Unger (Cambridge: Cambridge University Press, 2010). In many accounts, the idea of an American ecocide in

Vietnam has featured most dominantly; e.g., Stephen Macekura, *Of Limits and Growth: The Rise of Global Sustainable Development in the Twentieth Century* (Cambridge: Cambridge University Press, 2015), 120. For the increasingly difficult relationship between North and South, see Iris Borowy, *Defining Sustainable Development for Our Common Future: A History of the World Commission on Environment and Development (Brundtland Commission)* (London: Routledge, 2013); Caldwell and Weiland, *International Environmental Policy*; Cecilia Gowdy Wygant, "The United Nations Conference on the Human Environment: Formation, Significance and Political Challenges" (master's thesis, Texas Tech University, 2004); Macekura, *Of Limits and Growth*; Kai F. Hünemörder, *Die Frühgeschichte der globalen Umweltkrise und die Formierung der deutschen Umweltpolitik (1950–1973)* (Stuttgart: Franz Steiner Verlag, 2004); Johnson, *UNEP: The First 40 Years*.

10. John McCormick, *Reclaiming Paradise: The Global Environmental Movement* (Bloomington: Indiana University Press, 1991); "The Origins of the World Conservation Strategy," *Environmental Review* 10, no. 3 (1986); Macekura, *Of Limits and Growth*; John R. Weeks, *Population: An Introduction to Concepts and Issues* (Boston: Cengage Learning, 2014), 431.

11. E.g., FAO, *Report of the FAO Technical Conference on Marine Pollution and Its Effects on Living Resources and Fishing: Rome, 9–18 December 1970* (Rome: FAO, 1971); WHO, *National Environmental Health Programmes: Their Planning, Organization, and Administration; Report of a WHO Expert Committee* (Geneva: WHO, 1970); A. G. Forsdyke and WMO, *Meteorological Factors in Air Pollution* (Geneva: Secretariat of the World Meteorological Organization, 1970).

12. Peter H. Stott, "The World Heritage Convention and the National Park Service, 1962–1972," *The GWS Journal of Parks, Protected Areas & Cultural Sites* 28, no. 3 (2011): 285. For comprehensive accounts on political issues within the UN that in part related to the environment, also see Digambar Bhouraskar, *United Nations Development Aid: A Study in History and Politics* (New Delhi: Academic Foundation, 2007); Stanley Meisler, *United Nations: A History* (New York: Grove Press, 2011); Louis Emmerij, Richard Jolly, and Thomas G. Weiss, *Ahead of the Curve? UN Ideas and Global Challenges* (Bloomington: Indiana University Press, 2001).

13. Martin W. Holdgate et al., *The World Environment 1972–1982: A Report* (Dublin: Tycooly International, 1982), 7–8.

14. Rachel Carson, *Silent Spring* (Boston: Houghton Mifflin Harcourt, 1962); Patricia Lantier, *Rachel Carson: Fighting Pesticides and Other Chemical Pollutants* (St. Catharines: Crabtree Publishing Company, 2009).

15. Barry Commoner, *The Closing Circle: Man, Nature, and Technology* (New York: Knopf, 1971); Paul R. Ehrlich, *The Population Bomb* (New York: Buccaneer Books, 1968).

16. Paul Sabin, *The Bet: Paul Ehrlich, Julian Simon, and Our Gamble over Earth's Future* (New Haven: Yale University Press, 2014).

17. Donella H. Meadows et al., *The Limits to Growth: A Report for the Club of Rome's Project on the Predicament of Mankind* (New York: Universe Books, 1972).

18. Fernando Elichirigoity, *Planet Management: Limits to Growth, Computer Simulation, and the Emergence of Global Spaces* (Chicago: Northwestern University Press, 1999), 12.

19. See Sabine Höhler, *Spaceship Earth in the Environmental Age, 1960–1990* (London: Pickering & Chatto, 2015), 58. Putting the idea of a *Spaceship Earth* into its larger po-

litical context, Höhler has discussed how at the height of the Vietnam War this shared vessel seemed more vulnerable than ever.

20. Stott, "The World Heritage Convention and the National Park Service, 1962–1972," 18.

21. United Nations General Assembly, "2581(XXIV). Recommendation to Convene a United Nations Conference on the Human Environment" (1969), accessed 19 March 2019, http://www.un.org/ga/search/view_doc.asp?symbol=A/RES/2581%28XXIV%29.

22. Ehrlich, *The Population Bomb*; Commoner, *The Closing Circle*; Meadows et al., *The Limits to Growth*.

23. Barbara Ward and René J. Dubos, *Only One Earth: The Care and Maintenance of a Small Planet* (New York: W. W. Norton, 1972).

24. Maurice Strong in the preface of Ward and Dubos, *Only One Earth*, viii.

25. G. Evelyn Hutchinson et al., *The Biosphere* (New York: W. H. Freeman, 1970); UNESCO, ed., *Intergovernmental Conference of Experts on the Scientific Basis for Rational Use and Conservation of the Resources of the Biosphere, Paris, 4–13 September 1968: Recommendations* (Paris: UNESCO, 1968).

26. Other renowned correspondents included Giuseppe Montalenti, François Bourlière, Paul Duvigneaud, Otto Frankel, and Anna Medwecka-Kornas, all of whom had been involved in the IBP, and Mohamed Kassas, Frank Fraser Darling, Jean-Georges Baer, Maurits Mörzer-Bruijns, and Walery Goetel, all of whom were involved in IUCN and also partly in the MAB.

27. The section "The Discontinuities of Development," for example, contained this passage: "Rich and poor, developed and developing, industrial and pretechnological . . . all must inescapably share a single, vulnerable biosphere" Ward and Dubos, *Only One Earth*, 48.

28. Ibid., 112–14.

29. Another telling passage described the biosphere as such: "Alone in space, alone in its life-supporting systems, powered by inconceivable energies, mediating them to us through the most delicate adjustment, wayward, unlikely, unpredictable, but nourishing, enlivening, and enriching in the largest degree—is this not a precious home for all of us earthlings?" Ibid., 220, 42–43.

30. Ibid., 214.

31. Elichirigoity, *Planet Management*; Ward and Dubos, *Only One Earth*, 9–11.

32. Ward and Dubos, *Only One Earth*, 42–43. Elichirigoity has claimed that the cybernetic work by the American ecologist Edward O. Wilson has in fact influenced the work of science and technology studies scholars such as Donna Haraway, who is concerned with the modern machine-organism confluence. Donna J. Haraway, "In the Beginning Was the Word: The Genesis of Biological Theory," *Signs* 6, no. 3 (1981); Elichirigoity, *Planet Management*, 32.

33. Ward and Dubos, *Only One Earth*, 84.

34. Fuller (1895–1983), inventor of the geodesic dome and teacher at the Bauhaus-inspired Black Mountain College, designed infrastructure elements and dwellings from industrial designs and materials. Regarded as one of the key innovators of functional design of the second half of the twentieth century, Fuller was embraced by the counterculture movements of the postwar period, and functioned as one of the consultants for *Only One Earth*. For a detailed biography of Fuller, see Jonathon Keats, *You Belong to the Universe: Buckminster Fuller and the Future* (New York: Oxford University Press, 2016). For

a discussion of the links between systems engineering, space technology, and system biology, see Elichirigoity, *Planet Management*, 35–36.

35. R. Buckminster Fuller, *Operating Manual for Spaceship Earth* (Carbondale: Southern Illinois University Press, 1969), in Peder Anker, "Buckminster Fuller as Captain of Spaceship Earth," *Minerva* 45, no. 4 (2007): 426.

36. "The Ecological Colonization of Space," *Environmental History* (2005): 247. Höhler has linked this manipulation mentality to the Cold War atomic age, in which human action, in the form of a nuclear strike, for example, had in fact the potential to change—or end—planetary life at large. Höhler, *Spaceship Earth in the Environmental Age, 1960–1990*, 62.

37. "There is an inevitable and essential element of a redistribution of resources underlying the problem of the environment, just as there is a fundamental issue of social justice underlying every political order." Ward and Dubos, *Only One Earth*, 142.

38. Barbara Ward, *Spaceship Earth* (New York: Columbia University Press, 1966).

39. Ward and Dubos, *Only One Earth*, 122ff.

40. Daniel Speich Chassé, "Technical Internationalism and Economic Development at the Founding Moment of the UN System," in *International Organizations and Development, 1945–1990*, ed. Marc Frey, Sönke Kunkel, and Corinna R. Unger (London: Palgrave Macmillan, 2014), 23ff.

41. Ward and Dubos, *Only One Earth*, xix–xxv.

42. Vijay Prashad, *The Darker Nations: A People's History of the Third World* (New York: New Press, 2008), 180.

43. René Dubos, *Man Adapting* (New Haven: Yale University Press, 1965).

44. Ward and Dubos, *Only One Earth*, 174–80.

45. In the years leading up to the conference, conservationists at IUCN discussed this need for diplomatic skills and strategies to convince state actors at meetings and in their written exchanges, e.g., Gerardo Budowski to Harold Coolidge, 2 February 1970, HUA HJC: HUG (FP) 78.20, Box 2, Folder "IUCN-WWF and Related International Conservation 1970 IUCN HJC President–NY office of UN."

46. The American cultural anthropologists Margaret Mead, for instance, proposed the new science of ekistics, the study of human settlements, as one of the core disciplines to find more sustainable ways of living. See, e.g., Margaret Mead, "Anthropology and Ekistics," *Ekistics* 21, no. 123 (1966).

47. The scope of conservation competencies ascribed to organizations such as IUCN was determined at the conference by diplomatic discussions between nation-states on themes relating to international development. Recently, historians of science have started to develop the term of "science diplomacy." See, e.g., Simone Turchetti, *Greening the Alliance: The Diplomacy of NATO's Science and Environmental Initiatives* (Chicago: University of Chicago Press, 2018). While so far not a single definition has solidified, the term is useful in describing the ways in which past actors have appealed to science to engage in political or politicized debates, both with science as a subject for political discussion and scientific collaboration in the context of twentieth-century politics.

48. Holdgate, *The Green Web*, 104.

49. Harold Coolidge and Frank Nicholls, "Contact Note No 71/9: Natural Resources Development and Politics, 26 January 1971," HUA HJC: HUG (FP) 78.20, Box 7, Folder "IUCN-WWF and Related International Conservation 1971 IUCN International Rela-

tions: Other Unions, Etc. and UN"; Susan Reed to Harold Coolidge, 24 October 1971, HUA HJC: HUG (FP) 78.20, Box 7, Folder "IUCN-WWF and Related International Conservation 1971 ADM New York Office at UN"; Gerardo Budowski and Frank Nicholls, "Executive Board: Progress of IUCN Expansion Plans," HUA HJC: HUG (FP) 78.20 Box 1, Folder "IUCN-WWF and Related International Conservation 1970 IUCN-HJC President-Executive Board-May meeting."

50. "1972 Conference on 'Problems of Human Environment,'" 1969, HUA HJC: HUG (FP) 78.20, Box 3, Folder "IUCN-WWF and Related International Conservation 1970 IUCN International Relations: Conference on the Human Environment."

51. "Contact Note: Budowksi, Dasmann, Cerivsky, Hoffman, Warland, de Bonnaral contacted Maurice Strong on 29 September 1970," HUA HJC: HUG (FP) 78.20, Box 3, Folder "IUCN-WWF and Related International Conservation 1970 IUCN International Relations: Conference on the Human Environment" (2 of 2).

52. Holdgate, *The Green Web*, 112.

53. E.g., Rexmond D. Cochrane, *The National Academy of Sciences: The First Hundred Years, 1863–1963* (Washington, DC: National Academies Press, 1978), 488–90.

54. Raymond F. Dasmann, *Called By the Wild: The Autobiography of a Conservationist* (Berkeley: University of California Press, 2002), 173; IUCN, *Proceedings of the Tenth General Assembly, New Delhi 1969*, 27.

55. IUCN, ed., *Proceedings of the Eleventh General Assembly, Banff, Alberta, Canada 1972* (Morges: IUCN, 1972), 48ff; *IUCN Yearbook: Annual Report* (Morges: IUCN, 1972), 93–98.

56. IUCN, *Proceedings of the Tenth General Assembly, New Delhi 1969*, 132.

57. David Quammen, *The Song Of The Dodo: Island Biogeography in an Age of Extinctions* (New York: Random House, 2012), 14.

58. IUCN, ed., *Proceedings of the Eleventh Technical Meeting, New Delhi 1969* (Morges: IUCN, 1969), 169.

59. Gerardo Budowski, "More Power to IUCN's Elbow," published in the *New Scientist and Science Journal* on 15 April 1971, HUA HJC: HUG (FP) 78.20, Box 12, Folder "IUCN-WWF and Related International Conservation 1970 IUCN HJC President–NY office of UN."

60. William M. Adams, *Against Extinction: The Story of Conservation* (London: Earthscan; Fauna & Flora International, 2004), 170.

61. Paul B. Sears, "Review: The Careless Technology: Ecology and International Development. The Record of the Conference on the Ecological Aspects of International Development Convened by The Conservation Foundation and the Center for the Biology of Natural Systems, Washington University, December 8–11, 1968, Airlie House, Warrenton, Virginia. Mohammad Taghi Farvar and John P. Milton," *Quarterly Review of Biology* 48, no. 3 (1973): 520; Mohammad Taghi Farvar and John P. Milton, eds., *The Careless Technology: Ecology and International Development. The Record of the Conference on the Ecological Aspects of International Development, December 8–11, 1968, at Airlie House, Warrenton, Virginia* (Garden City, NY: Natural History Press, 1972).

62. E. Max Nicholson, "Statement from 5 March 1970." LSA EMN/IBP: Box 4, Folder "World Bank etc." The Adlai Stevenson Institute backed out after Maurice Strong decided not to take the leading role in the working group meeting. But in an inventive coup, Nicholson managed to bring the meeting to FAO.

63. Macekura, *Of Limits and Growth*, 178.
64. E. Max Nicholson, "Report of the Working Group on Ecology in International Economic Development," 1970, 2, LSA EMN/IBP: Box 4, Folder "World Bank etc."
65. Raymond F. Dasmann, John P. Milton, and Peter H. Freeman, *Ecological Principles for Economic Development* (New York: John Wiley & Sons, 1973), vi.
66. Raymond Dasmann to E. Max Nicholson, 26 October 1971, LSA EMN/IBP: Box 4, Folder "World Bank etc."
67. Also see Dasmann, Milton, and Freeman, *Ecological Principles for Economic Development*, vi, 3–5.
68. E.g., ibid., vii.
69. Also see Lynton K. Caldwell, *In Defense of Earth: International Protection of the Biosphere* (Bloomington: Indiana University Press, 1972); Caldwell, cited in Wendy R. Wertz, *Lynton Keith Caldwell: An Environmental Visionary and the National Environmental Policy Act* (Bloomington: Indiana University Press, 2014), 268; Milton, commenting in Farvar and Milton, *The Careless Technology*, 973.
70. Meadows et al., *The Limits to Growth*.
71. E.g., Harald Sioli, "Managing Natural Resources for Scientific, Education and Health Purposes," in *Proceedings of the Twelfth Technical Meeting, Banff, Alberta, Canada 1972*, ed. IUCN (Morges: IUCN, 1972), 224.
72. Garrett Hardin, *The Tragedy of the Commons* (Washington, DC: American Association for the Advancement of Science, 1968); IUCN, ed., *Proceedings of the Twelfth Technical Meeting, Banff, Alberta, Canada 1972* (Morges: IUCN, 1972).
73. McCormick, *Reclaiming Paradise: The Global Environmental Movement*, 78.
74. Dasmann in IUCN, *Proceedings of the Twelfth Technical Meeting, Banff Alberta, Canada 1972*, 133.
75. Coolidge in Farvar and Milton, *The Careless Technology*, 975.
76. IUCN, *Proceedings of the Tenth General Assembly, New Delhi 1969*, 133.
77. Gerardo Budowski to B. K. Steenberg, FAO Forestry Division, 19 February 1970, HUA HJC: HUG (FP) 78.20, Box 2, Folder "IUCN-WWF and Related International Conservation 1970 IUCN General: International Relations with International Unions and UN."
78. Harold Coolidge and Frank Nicholls, "Contact Note No 71/9: Natural Resources Development and Politics, 26 January," 1971, HUA HJC: HUG (FP) 78.20, Box 7, Folder "IUCN-WWF and Related International Conservation 1971 International Relations: Other Unions, Etc. and UN."
79. Helen Reus to Harold Coolidge, 23 September 1970, HUA HJC: HUG (FP) 78.20, Box 2, Folder "IUCN-WWF and Related International Conservation 1970 IUCN HJC President–NY office of UN."
80. Already in 1970, Helen Reus, a friend of Coolidge's at the UN headquarters, had asked him to find out about the UNESCO people involved in the preparations of the conference, so she could indicate who she deemed worth working with, and to compose a list of the most influential people who might be helping Strong (Helen Reus to Harold Coolidge, 23 September 1970, HUA HJC: HUG [FP] 78.20, Box 2, Folder "IUCN-WWF and Related International Conservation 1970 IUCN HJC President–NY office of UN").
81. Christina Buchhausen, "Weekly Report" to Frank Nicholls, 27 March 1970, HUA HJC:

HUG (FP) 78.20 Box 2, Folder "IUCN-WWF and Related International Conservation 1970 HJC President–NY office of UN."

82. Barrett Hollister, chairman of the NGO Committee to the Conference in which IUCN was much involved, to Harold Coolidge, 26 January 1970, HUA HJC: HUG (FP) 78.20, Box 2, Folder "IUCN-WWF and Related International Conservation 1970 IUCN HJC President–NY office of UN." With the financial assistance of the American Committee for International Wildlife protection, IUCN managed to arrange and fill a secretary position for the Committee of NGOs at the UN Headquarters in New York, where until 1971 the preparations for Stockholm were coordinated (Christina Buchhausen to Frank Nicholls, 13 May 1970, HUA HJC: HUG [FP] 78.20 Box 2, Folder "IUCN-WWF and Related International Conservation 1970 IUCN HJC President–NY office of UN").

83. Susan Reed to Harold Coolidge, 24 October 1971, HUA HJC: HUG (FP) 78.20, Box 7, Folder "IUCN-WWF and Related International Conservation 1971 ADM New York Office at UN."

84. Craig N. Murphy, *The United Nations Development Programme: A Better Way?* (Cambridge: Cambridge University Press, 2006), 60; also see Richard N. Gardner, *Sterling Dollar Diplomacy: Anglo-American Collaboration in the Reconstruction of Multilateral Trade* (Oxford: Clarendon Press, 1956).

85. Gerardo Budowski to Maurice Strong, 9 March 1971, HUA HJC: HUG (FP) 78.20, Box 8, Folder "IUCN-WWF and Related International Conservation 1971 IUCN-International Relations: UN Conference on the Human Environment-plans" (1 of 2).

86. Richard N. Gardner to Gerardo Budowski, 27 April 1971, HUA HJC: HUG (FP) 78.20, Box 7, Folder "IUCN-WWF and Related International Conservation 1971 IUCN ADM New York Office at UN."

87. Richard N. Gardner, "Paper Dealing with Post-Conference Institutional Arrangements," 1971, HUA HJC: HUG (FP) 78.20, Box 7, Folder "IUCN-WWF and Related International Conservation 1971 IUCN ADM New York Office at UN."

88. In spring 1971, IUCN tried to change their consultative status from category B to category A advisors, which would have given them a permanent representative status at all Stockholm-related meetings. This attempt failed due to South Africa's membership in IUCN. Decisions on the consultative status of non-UN organizations were made by the General Assembly on the basis of UN politics and membership. De Seynes, by now under secretary-general of ECOSOC, suggested that, instead, "the matter [IUCN's participation] could be dealt with on an ad hoc basis" (Philippe De Seynes, Harold Coolidge, Frank Nicholls, "Contact Note 71/8: IUCN Consultative Status with UNO," 1971, HUA HJC: HUG [FP] 78.20, Box 7, Folder "IUCN-WWF and Related International Conservation 1971 International Relations: other unions, etc. and UN").

89. Thorsten Schulz-Walden, *Anfänge globaler Umweltpolitik: Umweltsicherheit in der internationalen Politik (1969–1975)* (Munich: Oldenburg Verlag, 2013), 237.

90. Peter B. Stone, *Did We Save the Earth at Stockholm?* (Berkeley: Earth Island, 1973), 24–25.

91. "Information Note on Environmental Forum," 1972, HUA HJC: HUG (FP) 78.20, Box 12, Folder "IUCN-WWF and Related International Conservation 1972 IUCN UN Conference on the Human Environment, Stockholm, June 5–16." The Forum itself was put under the responsibility of the Swedish UN Association and the National Council of Swedish Youth and was sponsored largely by the Swedish government.

92. A management memo of November 1971 shows that the Forum's activities were to be strictly supervised ("Management Memo No 62, 22 November 1971," HUL PST II: Box 15, Folder 130). Also see Christina Buchhausen, "Bimonthly Report" to Frank Nicholls, 28 June 1971, HUA HJC: HUG (FP) 78.20, Box 7, Folder "IUCN-WWF and Related International Conservation 1971 IUCN ADM New Work Office at UN."

93. For the composition of the Forum, see Macekura, *Of Limits and Growth*, 120.

94. "Statement of the NGOs to the Plenary Session of UNCHE, 12 June 1972," HUA HJC: HUG (FP) 78.20, Box 12, Folder "IUCN-WWF and Related International Conservation 1972 IUCN UN Conference on the Human Environment, Stockholm, June 5–16."

95. Terri Aaronson, "World Priorities," *Environment: Science and Policy for Sustainable Development* 14, no. 6 (1972): 4.

96. United Nations Conference on the Human Environment, "Declaration on the Human Environment, A/CONF.48/PC.11/Add.42, Stockholm, 16 June 1972," (1972), accessed 19 March 2019, http://webarchive.loc.gov/all/20150314024203/http%3A//www.unep .org/Documents.Multilingual/Default.asp?documentid%3D97%26articleid%3D1503.

97. See Macekura, *Of Limits and Growth*, 125; Indira Gandhi, *Man and His Environment* (New Delhi: India Book Centre, 1973).

98. Wade Rowland, *The Plot to Save the World: The Life and Times of the Stockholm Conference on the Human Environment* (Madison: Clarke, Irwin, 1973), 36.

99. Commission on International Development and Lester B. Pearson, *Partners in Development: Report* (New York: Praeger, 1969); Matthias Schmelzer, "The Club of Rome to Help the Poor? The OECD, "Development," and the Hegemony of Donor Countries," in *International Organizations and Development, 1945–1990*, ed. Frey, Kunkel, and Unger, 205.

100. Sönke Kunkel, "Contesting Globalization: The United Nations Conference on Trade and Development and the Transnationalization of Sovereignty," in *International Organizations and Development, 1945–1990*, ed. Frey, Kunkel, and Unger, 243–48.

101. "Bimonthly Report," to Richard N. Gardner, 10 October 1971, HUA HJC: HUG (FP) 78.20, Box 8, Folder "IUCN-WWF and Related International Conservation 1971 IUCN-International Relations: UN Conference on the Human Environment-plans" (2 of 2).

102. Ronald Bailey, "Who Is Maurice Strong?," *National Review,* 1 September 1997; Rowland, *The Plot to Save the World: The Life and Times of the Stockholm Conference on the Human Environment*; Maurice F. Strong, *Where on Earth Are We Going?* (Toronto: Knopf Canada, 2000).

103. E.g., *Where on Earth Are We Going?*, 87ff; Bailey, "Who Is Maurice Strong?"; Sam Roberts, "Maurice Strong, Environmental Champion, Dies at 86," *New York Times*, 1 December 2015, http://www.nytimes.com/2015/12/02/world/americas/maurice-strong-environmental-champion-dies-at-86.html?_r=0.

104. E.g., Maurice F. Strong, *Canada's Assistance to Developing Nations* (Vienna: Vienna Institute for Development, 1970); also see Emmerij, Jolly, and Weiss, *Ahead of the Curve?*, 48–59.

105. With Monod's help, Strong had been in line for a junior clerical role for the Palestine Commission in late 1948. Yet, a few months prior to the planned appointment, the Swedish mediator in the Arab-Israeli conflict, Folke Bernadotte, was assassinated and the mission consequentially restructured (Strong, *Where on Earth Are We Going?*, 73).

106. Rowland, *The Plot to Save the World*, 35–37.
107. Maurice Strong, "Interview with 'Delegates World,'" 1974, transcript, HUL MFS: Box 27, Folder 267 "IV Interviews."
108. "The Crisis of Our Environment and the Quality of Life, Address by Maurice F. Strong, Secretary-General, United Nations Conference on the Human Environment at the Fourth Session of the UN Commission on International Trade Law, 17 November 1971," HUL MFS: Box 28, Folder 281.
109. Ibid., 9.
110. "Reference Materials," n.d., HUL MFS: Box 28, Folder 283.
111. "The Crisis of Our Environment and the Quality of Life, Address by Maurice F. Strong.
112. "Interview with M. F. Strong," Interviewer: B. Martin from the Canadian Broadcasting Corporation, Toronto, 1975, transcript, HUL MFS: Box 27, Folder 267 "IV Interviews."
113. "Note for the Files," 12 April 1971, HUL PST II: Box 11, Folder "99 GA Conf. 48."
114. McCormick, *Reclaiming Paradise*, 78.
115. UN and ECOSOC, *Science and Technology for Development: Proposals for the Second United Nations Development Decade; Report of the Advisory Committee on the Application of Science Technology to Development* (New York: UN, 1970), 6.
116. Robert McNamara, "Address to UNCHE, 8 June 1972," HUL PST II: Box 11, Folder 99 "GA Conf. 48."
117. Robert Gillette, "The Limits to Growth: Hard Sell for a Computer View of Doomsday," *Science* 175, no. 4026 (1972): 1091; also see Christopher Freeman, "Prometheus Unbound," *Futures* 16, no. 5 (1984); Annemieke J. Roobeek, "The Crisis in Fordism and the Rise of a New Technological Paradigm," *Futures* 19, no. 2 (1987): 137.
118. Maurice Strong, "Opening Statement to the Stockholm Conference," 1972, HUL MFS: Box 28, Folder 278.
119. E.g., Maurice Strong, "Notes on Scientific and Conceptual Framework, 30 October 1970," HUL MFS: Box 29, Folder 291 "IV Scientific and Conceptual Framework."
120. "Environmental Fact File, 23 May 1972," HUL MFS: Box 29, Folder 291 "IV Scientific and Conceptual Framework."
121. John McHale, *The Ecological Context* (New York: G. Braziller, 1970), 147.
122. John McHale, *The Future of the Future* (New York: G. Braziller, 1969).
123. Ibid., 5.
124. Ibid., 11.
125. Ibid., 96.
126. Maurice Strong, "Opening Statement to the Stockholm Conference," 1972, 14–15, original emphasis, HUL MFS: Box 28, Folder 278.
127. "Panel Discussion on the International Organizational Implications," 1970, HUL MFS: Box 29, Folder 291 "IV Scientific and Conceptual Framework."
128. Strong's paper from 23 September 1971, titled "Outline and Main Content of the Report on Institutional Implications of the Action Proposal," written for the Stockholm Conference, stressed the need for the "UN machinery" to give a "strong, well-articulated and persuasive exposé of the reasons why [environmental] multilateral co-operation should be implemented within the UN framework. . . . Reference should be made in this context to the International Development Strategy." See HUL PST II: Box 14, Folder 121 "Conf 48 Misc."; "Notes on Possible Alternatives for Post-Stock-

holm Organizational Arrangements," 1971, HUL PST I: Box 15, Folder 131 "Conf. 48 Panel of Experts."

129. UN, "An Action Plan for the Human Environment, A/CONF.48/14/Rev.1, Stockholm, 16 June 1972," (1972), http://www.un-documents.net/aconf48-14r1.pdf.

130. McCormick, *Reclaiming Paradise*, 106–13.

131. Bob Reinalda, *Routledge History of International Organizations: From 1815 to the Present Day* (New York: Routledge, 2009), 516.

132. "UNEP Governing Council. Second Session, Nairobi, March 11–22, 1974. Item 8a of the Provisional Agenda," HUL PST II: Box 65, Folder 611.

133. Ibid.

134. McCormick, *Reclaiming Paradise*.

135. Changed in 1973 to International Institute for Environment and Development (IIED).

136. Thomas W. Wilson to Maurice Strong, 16 July 1973, HUL MFS: Box 33, Folder 334 "III Aspen 'Outer Limits.'"

137. John McHale and Magda McHale, "Special Consultation for Maurice Strong: Executive Director, United Nations Environmental Program, August 19–25, 1973 at Aspen, Colorado," HUL MFS: Box 33, Folder 334 "III Aspen 'Outer Limits.'"

138. "Outline—The Institutional Aspect of the Environment, 20 May 1973," HUL MFS: Box 28, Folder 281 "IV Reference Materials."

139. IUCN, *Proceedings of the Eleventh General Assembly, Banff, Alberta, Canada 1972*, 51.

140. Peter S. Thacher, "What's Happened since Stockholm," *Environmental Science & Technology* 8, no. 3 (1974): 214.

141. Ibid.

142. Holdgate et al., *The World Environment 1972–1982*, 9.

143. Recommendations 20 and 24 of UN, "Report of the United Nations Conference on the Human Environment. Stockholm, 5–16 June 1972," accessed 23 March 2019, http://www.un-documents.net/aconf48-14r1.pdf, 9, 11.

144. Thacher, "What's Happened since Stockholm."

145. Thacher, "Next Steps," to Strong, 27 June 1972, HUL PST II: Box 15, Folder 130.

146. United Nations Conference on the Human Environment, "Declaration on the Human Environment, A/CONF.48/PC.11/Add.42, Stockholm, 16 June 1972," (1972), accessed 19 March 2019, http://webarchive.loc.gov/all/20150314024203/http%3A//www.unep.org/Documents.Multilingual/Default.asp?documentid%3D97%26articleid%3D1503, see principle 4.

147. Whereas IUCN had tried to modernize and reform away from older preservation-based ideas of conservation, within its structure, the Species Survival Commission and the Commission for National Parks remained two of the oldest, largest, and most active commissions in the organization. See Martha Coolidge to Lee Talbot, 29 April 1977, HUA HJC: HUG [FP] 78.16, Box 7 ABC Series, Folder "C Colleagues and Friends, Talbot, Dr Lee M., wife Martha."

148. Stott, "The World Heritage Convention and the National Park Service, 1962–1972."

149. John Perry, assistant director at the National Zoological Park, Smithsonian Institution, to Fred Evenden, secretary treasurer of the American Committee for International Wildlife Protection, Inc., The Wildlife Society, and to Lee Talbot, 22 September 1971, HUA HJC: HUG (FP) 78.20, Box 11, Folder "IUCN-WWF and Related International Conservation 1972 ACIC American Committee on International Wildlife Protection."

150. IUCN, *Proceedings of the Eleventh General Assembly, Banff, Alberta, Canada 1972*, 16.

151. Ibid.; IUCN, *IUCN Yearbook: Annual Report.*

152. IUCN, ed., *Proceedings of the Twelfth General Assembly, Kinshasa 1975* (Morges: IUCN, 1976), 45.

153. UNESCO received $15,900 for an overview of ecosystems, sites and samples, national parks, and reserves, and another $233,000 for biosphere reserve establishment and survey (KIT: "UNEP Annual Review 1975").

154. Peter Scott, "Conservation—The Task of Our Time, Budget Requirements, 15 November 1974," CUSC PMS: NCUACS 87/8/99, Box C.1175.

155. Martha Coolidge to Lee Talbot, 29 April 1977, HUA HJC: HUG (FP) 78.16, Box 7 ABC Series, Folder "C Colleagues and Friends, Talbot, Dr Lee M., wife Martha."

156. Lynton K. Caldwell, "Concepts in Development of International Environmental Policies," in *Proceedings of the Twelfth Technical Meeting, Banff, Alberta, Canada 1972*, ed. IUCN, 95–96.

157. In the late 1960s, Caldwell had been involved as a scientific adviser in the foundation of NEPA.

158. The German historian Sönke Kunkel has used this expression to describe the "open space where compliance is produced and negotiated through carrying means that encompass not only financial or military resources, but also moral pressure, discourse power, knowledge or the mobilization of public outrage." Sönke Kunkel in *International Organizations and Development, 1945–1990*, ed. Frey, Kunkel, and Unger, 242.

 CHAPTER 4

The Fault Lines in the World Conservation Strategy, 1975–1980

Introduction

The years following the Stockholm Conference were some turbulent ones for international environmental policymaking. By the mid-1970s, the political tensions between so-called developed and undeveloped countries were stirred up by heavy droughts in the African Sahel region and an oil embargo by Arab states against importing Western countries. These food and energy crises, followed by rising demands by governments of the global South for a New International Economic Order, seemed to require a rethinking of the ways the United Nations (UN) were dealing with the limits to natural resources. As socioeconomic solutions seemed insufficient, scholarly works that addressed the biological prerequisites for economic development were reconsidered. In 1975, representatives of the United Nations Environmental Program (UNEP), the United Nations Educational, Scientific, and Cultural Organization (UNESCO), and the UN's Food and Agricultural Organization (FAO), invited scientists from the International Union for Conservation of Nature and Natural Resources (IUCN) to the Ecosystem Conservation Group. IUCN conservationists were not only asked to join this new inter-organizational alliance; UNEP's governing council also commissioned them to draft a document called *World Conservation Strategy*, which was to guide the work of international organizations and national governments on conserving natural resources around the globe.

This chapter is concerned with the consequences of this seemingly new turn of UN scientists and policymakers to nature conservation. After the disappointments of the Stockholm Conference, where nature conservation had been confined to traditional themes such as the protection of threatened wildlife and national parks, leading members at IUCN regarded the task of developing a shared *Strategy* as a new opportunity to make broadly defined conservation a more essential part of environmental politics at the UN level. This chapter asks what the decision to draft a *World Conservation Strategy* meant for the scientific content of and the implementation strategies for global nature conservation, as well as for IUCN members' role in international environmental policymaking. Further, it looks at the kinds of environmental coop-eration between IUCN and UNEP that the initiative brought about, and how the final program outlined in the *Strategy* related to the ecosystem approach that IUCN conservationists had been developing over the years.

Historians have predominantly interpreted the *Strategy* as the first UN-related document to firmly conjoin the environment and the economy, coin-ing the term *sustainable development* in the process.[1] In fact, the *Strategy* has been celebrated as a major turn toward integrating environmental concerns into resource and development policies. This turn entailed a move away from a posteriori measures, remedying environmental harm only after soils had become unproductive or species threatened by extinction. Instead, the story goes, the *Strategy* presented a move toward environmental protection a priori, by integrating projections on potential environmental pollution or resource overexploitation early in land or agricultural development projects.[2] Based on a similar interpretation, long-term member of IUCN Martin Holdgate has called the *Strategy* the organization's "peak of achievement" and the "most im-portant single contribution in the history of IUCN."[3]

The history of the concept of sustainable development and of the different organizations involved in bringing it into international environmental discus-sions is important, as the concept remains central to environmental policy-making even today. The focus here, however, will be less on the negotiation of a shared discourse, and more on the tensions that defined the intellectual underpinning of the *Strategy*. By examining the scientific arguments by lead-ing IUCN members in the drafts of the *Strategy*, the scientific sourcebooks underlying these, and the critical comments by UNEP consultants, I argue that the *World Conservation Strategy* did not lead to a new, shared environmental vision. Rather than producing one unifying ideology, UNEP and IUCN devel-oped two distinct agendas for balancing natural and social aspects of environ-mental degradation. Although members of both organizations dedicated their work to the idea of sustainable development and the obligation to fulfill *ba-sic human needs*, they developed different ecological arguments and opposite ways of measuring the societal value of nature, and they pursued their work through two separate scientific networks.

Ways in which conservation and touristic activities or processes have been steered to generate economic value of the natural environment have gener-ated much literature in recent years.[4] While these accounts have focused on the economies of wildlife, the notions of value covered by this chapter are not limited to direct economic advancements. Ideas on the value of nature as dis-cussed here included a temporal dimension, in which long-term or potential benefits from healthy natural systems—ideas now linked to sustainability—play a role next to the market value of particular natural resources, such as wood, ore, or clean water. On the one hand, a number of IUCN conservationists, drawing on a new discourse in the scientific field of island biogeography, argued for a global network of samples of ecosystems and the biological diversity they contained. This way, the stability of the biosphere, which according to them underlay all attempts to improve the human condition, could be secured. In this line of argumentation, nature was assigned an original ecological value that was of greater significance than any direct economic revenue. On the other hand, UNEP scientists turned to new ecological theories on resilience and methods for environmental impact assessment. This approach seemed to allow for a decentralized integration of environmental factors, tailor-made for individual development projects. At the same time, the approach by UNEP scientists encouraged the integration of natural resources and natural processes into the market economy, quite opposite to IUCN conservationists' emphasis on the natural or ecological value of biological diversity. The tensions between their approaches and networks remain visible even today.

Environmental Limits

It might come as a surprise that UNEP's governing council put discussions on the biological limits and the conservation of natural resources back on the agenda during the mid-1970s. After all, the Stockholm Conference had set a brisk end to the plans of IUCN Director General Budowski and his peers to promote the broad relevance of conservation knowledge and to profile IUCN as the scientific nongovernmental organization (NGO) with the most expertise in global environmental questions. The idea that ecosystem ecology could function as the scientific underpinning to streamline the UN's environmental work had gone unheeded. However, the Stockholm Conference had not assuaged the manifold concerns about an imminent global environmental crisis. Instead, continuing debates on population growth and the overexploitation of natural resources contributed to a climate of instability within the international institutions that had been involved in the conference.[5]

In Stockholm, much of the socioeconomic remedies proposed for environmental problems had been based on the idea of a stabilized economy and relatively predictable markets, rooted in the observation that the 1960s had been a decade of substantial economic growth. Yet the early 1970s seemed to indicate a decline of growth rates, especially in the developed world. This was further exacerbated by a transnational food and trade crisis shared between less developed and developed countries. In 1972 and again in 1974, cereal and rice yields failed due to unfavorable weather events in many crop-producing regions, leading to an overall increase in global food prices. Two heavy droughts in Ethiopia and the Sahel countries further worsened the food shortage in many

African countries. Confronted with its first major crisis, the United Nations Development Program (UNDP) was struggling to perform, especially as U.S. funding was in decline.[6] Problems for the global South were made worse after U.S. President Richard Nixon withdrew the fixed gold convertibility of the dollar in 1971. As a result of this measure, intended to improve the United States' foreign trade position, markets in the global South saw their export prices fall.[7]

A number of country representatives from the global South to the UN, to a large degree depending on the export of primary resources, began to protest that the existing international aid system was no longer working.[8] These protests attained more leverage in 1973, when oil-producing countries of the Arab Organization of the Petroleum Exporting Countries (A-OPEC) reacted to a U.S. intervention in the Israel conflict by demanding higher prices for crude oil.[9] The oil embargo caused a substantial increase of global oil prices and shortages in the United States and Western Europe. In the short run, the embargo revealed the vulnerability and the dependency of industrialized countries on resource supply from other parts of the globe.[10] In the long run, however, the oil crisis raised the prices for agricultural fertilizers and pesticides, contributing to a loss of further yields, which mostly affected parts of the less developed world, especially Africa and the Pacific region.[11] These transnational developments made natural resources, their development, and their distribution highly charged topic areas for international political deliberations.

Thus, in the years following the Stockholm Conference, the global dependencies of societies on natural resources and, in turn, the societal consequences of breaching environmental limits seemed an incontrovertible fact. At the same time, not all geographical regions seemed to suffer from these global dependencies in equal ways. The gap between incomes per capita in the global South and North continued to widen.[12] It was not long before a group of newly independent countries, since 1964 united in the Group of 77, a coalition of developing countries, demanded a radical overhaul of the international economic system. Their discontent came from the structures engrained in many of the organizations that regulated financial relations between UN member states, such as the World Bank, the International Monetary Fund (IMF), or the General Agreement on Tariff and Trade (GATT). The Group of 77, by then holding a majority within the UN, particularly used the forum of the United Nations Conference on Trade and Development (UNCTAD) for their claims. Here, members of the Group of 77 called for a New International Economic Order. This new order was to replace the existing set of rules for international trade, which barely distinguished between affluent and less affluent national economies, with two sets of rules that differentiated trade agreements between countries from the global North from those between developed and less developed countries.[13] Although regarded critically by some powers in the global North, within the UN the sentiment grew that global inequalities regarding the access to natural resources demanded further revisions of international trade and aid.[14]

It was in this moment of rethinking international economic relations that UNEP's governing council, the new coordinating body for UN environmental affairs, assumed office. If UNEP scientists wanted to play an authoritative role within the UN, they had to do justice to calls for both further economic reform and additional biological research on the environmental limits to growth. UNEP's new executive director, Maurice Strong, had continuously "preached" at Stockholm "the compatibility of development and environmental quality," and, as a former American consultant and observer to the Stockholm Conference reported, this rhetoric had begun to "rub off on the delegates."[15] Scientists and policymakers at UNEP mimicked this language. During UNEP's first meeting in 1973, the members of the governing council began to adapt an agenda that acknowledged two sets of problems: first, those pertaining to the socioeconomic sphere, and, second, those relating to the biosphere.[16] At the same meeting, Strong suggested that in order to do justice to these two objectives, two research programs had to be designed and coordinated by UNEP to steer the work of other agencies within the UN. A first attempt, which he called "the Outer Limits Program," was to deal with the natural balance of the environment itself, a second one, "the Inner Limits Program," concerned the social causes and the economic consequences of environmental degradation.[17]

At first glance, Strong's suggestion might look similar to the idea that ecological principles should underlie all decisions on environmental planning and development, propagated by Dasmann or Budowski as discussed in the previous chapter, or might even remind one of Max Nicholson's systems thinking during the International Biological Program (IBP).[18] Strong's approach, however, was quite a different one. It served to address both the natural and the social limits to environmental degradation, but, rather than finding shared methods to manage both sets of limits at once, Strong treated them as two distinct domains. In this double venture, the "outer limits" could not be breeched, while the "inner limits" had to be guaranteed.[19] One can visualize the differences by thinking of two circles, one enclosing the other (figure 4.1).

In this visualization, the outer ring—limits to natural resources—is fixed. The area within the inner ring—basic human needs—must not shrink under a certain size. Development, then, is possible in the strip between the two rings, and, depending on the size of the human population, leads to the encroachment of the inner on the outer ring. The idea was to find ways to come close to the outer ring without breeching it. In short, this was not one working system, but two systems working against each another.

In order to devise ways in which the two systems could be balanced, in 1973 Strong and UNEP's deputy executive director, the Egyptian microbiologist Mostafa Tolba, first convened a scientific meeting to discuss a potential Outer

Figure 4.1. The "inner" and the "outer" limits to the environment. My visualization of Strong's rhetoric, based on "Excerpt from Mr. Maurice Strong's Introductory Statement to UNEP's Governing Council, 12 June 1973," HUL MFS: Box 33, Folder 334 "III Aspen 'Outer Limits.'"

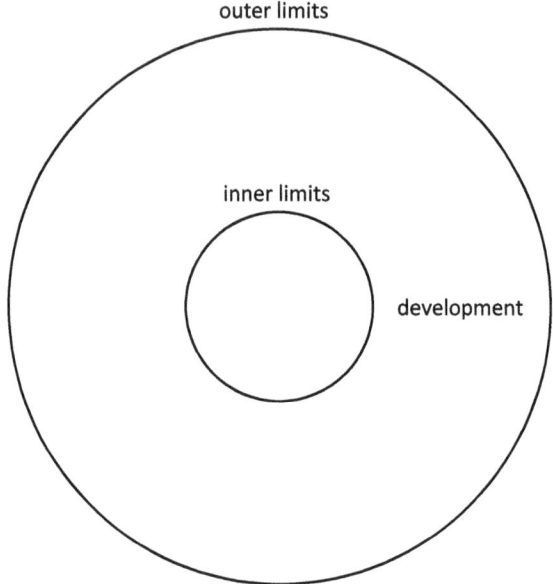

Limits Program for UNEP at the North American Aspen Institute for Integrative Studies. The meeting was attended by many of Strong's former consultants from the time of the preparations for the Stockholm Conference. Among them were the systems thinker John McHale; *Only One Earth* coauthor René Dubos; Carroll Wilson from the MIT systems group; Richard N. Gardner, who had represented IUCN at the Stockholm Conference; Thomas Malone, working for ICSU's Scientific Committee on Problems of the Environment (SCOPE); and Mohamed Kassas, former IBP and IUCN member, at the time working for both SCOPE and UNESCO. Themes considered for the program were based on the idea of finite natural resources. For instance, topics they dealt with related to global protein supply, additional energy sources, and the safety of problems of atomic energy. Discussed, too, were the carrying capacities of land and water ecosystems; questions on "thresholds," "triggers," and the "irreversibility" of human-made changes; population issues, including fertility and contraception; and constraints of limited global resources—in this way reintroducing themes that had received less attention in Stockholm.[20]

Strong hoped that his broad agenda for the Outer Limits Program would help connect the natural science projects on aspects of a broadly defined environment that already existed within the UN and bring them under UNEP's supervision. Within the different UN agencies, work related to the topics of natural resources in the new context of the environment had gained new attention since the Stockholm Conference. By 1973, UNESCO ecologists had developed two initiatives concerned with protecting the biosphere, the Man and the Biosphere Program (MAB) and the World Heritage Convention. Moreover, since the Sahel food crisis, scientists linked to UNESCO and FAO had placed new emphasis on the prevention of desertification and on soil conservation in arid regions.[21] FAO scientists especially were engaged in extensive research programs on land and marine resources, and increasingly in new projects on agricultural improvement through genetic modifications, for example in Thailand and the Philippines.[22]

In October 1974, the Inner Limits Program, pertaining to society and economy, also attained more shape.[23] At a conference on the "Patterns of Resource Use, Environment and Development Strategies" held in Cocoyoc, Mexico, under the joint auspices of UNCTAD and UNEP, natural resource problems were discussed that did not relate to "absolute physical shortage, but [to] . . . economic and social maldistribution and usage," as the conference declaration explained.[24] The Cocoyoc Conference issued a declaration in which the "inner limits" were defined in an anthropocentric way as the satisfaction of "basic human needs."[25] The expression was taken from a speech held in 1972 by Robert McNamara, president of the World Bank. In the following year, the idea of basic human needs developed alongside international economic reforms that focused on securing minimum standards of living conditions and wellbeing worldwide. This new approach originated out of a strong resentment of intergovernmental aid treaties that benefited only the political elites in developing countries but neglected the welfare of the majority of the population. Supporters of the idea of basic human needs included not only economists at the World Bank but also the aforementioned Aspen Institute, the International Institute for Environmental and Development (IIED, the formerly IIAE) run by Barbara Ward, and ECOSOC, headed by De Seynes. In 1974, it was taken up by UNEP under Strong and Tolba.[26]

With the two programs in place, Strong and Tolba hoped to put UNEP into a position to coordinate the work of UN agencies both on social issues, such as environmental justice, and on scientific research on biological problems related to the potential resource limits of the planet, desertification, or the carrying capacity of ecosystems. Yet, by itself, the new acknowledgement of two types of problems did not secure the position Strong wanted for UNEP. In particular, his objective to coordinate the existing environmental research and projects led by scientists within the UN agencies through the channels of UNEP was not immediately supported by the members of UNESCO and FAO. With their own projects in place and no prospect of additional funding, there was little incentive for them to subordinate their work under the supervision of UNEP's governing council. FAO officials especially had reacted with indignation when UNEP took over leadership in the drafting of a UN Action Plan against the pollution of the Mediterranean Sea in 1974.[27] Strong and Peter Thacher, a North American expert in geographic information systems from the UNEP Geneva office, had subsequently commissioned UN-external scientists at SCOPE to work out the Action Plan, which had complicated the relationship with FAO even more.[28]

Members of UNEP would have had a hard time coordinating the diverse international initiatives on the environment even if their headquarters had been located close to the seats of FAO or UNESCO, in Rome or Paris. As it was, however, the location of UNEP's seat in Nairobi made it even more difficult for Strong to liaise with the different UN agencies.[29] In fact, the strategic decision for a headquarters in Kenya had left the young organization and their new executive director virtually isolated for the first few years. Strong later recalled that in 1973 it was only him and a resolution by the UN General Assembly that supported the establishment of UNEP in Nairobi.[30] By 1975, without much support for the Outer Limits Program, there was even talk of simply merging UNEP with ECOSOC and UNCTAD.[31] At the same time, suggestions were made within the UN's executive ranks to make UNEP a real agency of the developing world by replacing the Canadian Strong with a representative from the global South.[32]

Thus, when in 1975 Strong summoned together representatives of UNESCO, FAO, and members of IUCN to form a new environmental task force, the Ecosystem Conservation Group, it was to legitimize the very existence of UNEP by gaining additional support for its twofold program. A 1975 article by Strong for the scientific journal *Environmental Conservation* shows that he hoped the new group would bring more UN agencies on UNEP's side by devising "a new system more capable of meeting the inner limits of basic human needs for all the world's people," but "without violating the outer limits of the planet's resources and environment."[33] This attitude he called a "new commitment to conservation," giving the Ecosystem Conservation Group its name.[34]

Unlike the members of FAO and UNESCO, leading conservationists at IUCN responded positively to Strong's suggested interorganizational alignment.[35] In 1975, the *IUCN Bulletin,* the main means of the executive board to communicate with IUCN's numerous members, spoke of a new boost for conservation, which logically would involve a bigger role for IUCN and a closer cooperation with UNEP and the other UN specialized agencies.[36] For many IUCN members, the prospect of additional recognition by UNEP came at a convenient moment. For IUCN, the years of 1975 and 1976 marked a period of internal ruptures. The organization's different commissions had been growing farther and farther apart.[37] The overall budget was tight, and the management was still that of a small voluntary organization. By the mid-1970s, a form of pragmatism had set in and replaced the ambitious plans that Nicholson or Budowski had pursued during the previous years to promote IUCN's ecological expertise in development, resource, and population management.[38] The bulk of IUCN's advisory and publication work was carried by the Commission on National Parks and Protected Areas and the Species Survival Commission (SSC), which, thanks to their widely ramified international networks, were best funded and best known in many parts of the world.[39] Yet, the independence of these commissions undermined the actual power of the IUCN headquarters.

Overall, IUCN members welcomed UNEP's initiative at a time when their organization's image as an ecological research agency and the main scientific authority in conservation matters was crumbling. By 1977, IUCN's former showpiece commission, the Commission on Ecology, according to its own chairman, John Ovington, had "run out of steam" and was financially troubled, as he explained at a meeting of the IUCN commission chairs in early 1978.[40] This had as much to do with a general lack of funding as with new international initiatives that had outstripped IUCN's research on ecology-based conservation and resource management. In September 1974, for instance, the International Association for Ecology had organized a First International Congress of Ecology in The Hague, which had brought together members of diverse scientific organizations, including scientists from UNESCO, FAO, and SCOPE. IUCN was only sparsely represented.[41]

Within the UN, furthermore, agencies such as FAO were in the process of developing their own scientific approaches to conservation, with a new focus on agricultural genetics research.[42] Genetic diversity had become part of the larger task field of natural resource conservation after the United Nations Stockholm Conference of 1972. A "whole raft" of recommendations on the need to protect "genetic pools" had been decided upon in Stockholm.[43] These had promoted the establishment of national and regional seed banks for genetic conservation and experimentation to fight hunger and malnutrition in the so-called developing world. Immediately after the Stockholm Conference, an "emergency programme" on plant gene conservation in national and regional seed banks had been assigned to FAO and parts of UNESCO's MAB.[44] These alternative conservation approaches *ex situ,* outside of the natural environment, required neither the guidance of IUCN conservationists nor the ecosystem approach they promoted.

By the mid-1970s, IUCN's conservation advice was in little demand. IUCN was not consulted when, in 1974, the UN organized a series of political summits on topics related to the limits of natural resources, including a World Food Conference under the aegis of FAO and a UN World Population Conference in Bucharest. Neither did IUCN members play an important role when a year later, in 1975, the UN convened its third Conference on the Laws of the Sea and the first UN Conference on Desertification. Dasmann wrote in the

Bulletin that by discussing these topics at separate conferences and with little involvement of IUCN, UN policymakers clearly ignored the ecological rules of the biosphere that IUCN had been defending.[45] At the same time, IUCN scientists feared that UN-sponsored programs in both FAO and UNESCO could jeopardize the status of IUCN as the main international scientific conservation organization. For their work relating to the environment, each of the two organizations received four times as much funding from UNEP as did IUCN.[46]

Under the pressure of external competition, frictions also emerged within the IUCN commissions and the secretariat, eventually triggering significant reforms in the organization's administrative setup.[47] In 1975, conflicts flared in the General Assembly between conservationists who were in favor of a more active management of natural resources and those preferring noninvasive forms of nature protection.[48] Discontent with the new appointments made at the General Assembly, Budowski and Director General Frank Nicholls left IUCN.[49] At the IUCN General Assembly in 1977, further administrative changes were introduced. One of the most significant ones was that Strong himself moved into IUCN's executive ranks as the new chairman of the Bureau, an intermediate body representing the council, which in turn had replaced the executive board.[50] From inside IUCN, Strong set in motion several additional institutional changes.

Strong had left UNEP in 1975 for a position at the oil company Petro Canada. After hierarchical discrepancies with UN Secretary Kurt Waldheim, some of Strong's colleagues at UNEP suggested that he was a better entrepreneur than a manager. At UNEP, Tolba had taken Strong's place, which satisfied those who wanted UNEP to be run by a representative of a country from the global South, as well as those who wished for a committed scientific expert.[51] Strong's migration from UNEP to IUCN marked the beginning of a new period of cooperation between the two organizations. With Strong, new funding began to stream in for IUCN.[52] In several ways, Strong saw IUCN as a tool to link environmentally minded NGOs and governmental organizations to UNEP, while many members of IUCN saw in Strong an administrator that would be good to have on board.[53] A new contract between UNEP and IUCN was negotiated in the summer of 1977, putting IUCN firmly in charge of the *World Conservation Strategy*. Poore, IUCN's scientific director, retained a crucial function as scientific adviser when in 1977 the development of the *Strategy* became one of IUCN's most important tasks. In fact, Poore hoped that the work on the *Strategy* would create new shared momentum among the more than seven hundred scientists affiliated with IUCN, while strengthening the profiling of IUCN's work within the Ecosystem Conservation Group.[54] The drafts of the *World Conservation Strategy* reveal the intention of leading IUCN conservationists to underpin their position vis-à-vis the other member organizations of the Ecosystem Conservation Group: UNEP, UNESCO, and FAO.

Ecological Stability through Biological Diversity

Between 1977 and 1980, the *Strategy* went through three internal drafts. It was mainly Robert Allen, a science writer who in 1972 had been a coauthor of the *Blueprint to Survival,* and IUCN's senior ecologist Poore who were involved in bringing together the material from which the *Strategy* was composed.[55] In addition to the extensive work of the Species Survival Commission's *Red List of Threatened Species* and the Committee for National Parks and Protected Areas' *Lists of National Parks,* Allen consulted more than four hundred of IUCN's members on conservation priorities, which were used for a first draft.[56] In January 1978, the first draft was published and sent for comments to more than one thousand IUCN members. Poore left IUCN's administration in 1978, after Strong had appointed another former UNEP member, the Canadian microbiologist David Munro, as IUCN's permanent director general.[57] Nevertheless, Poore and his scientific ideas remained pivotal for the composition of the different drafts of the *Strategy.*[58] A second draft, finalized six months later, came with six short sourcebooks of approximately twenty pages each. These sourcebooks presented the main scientific framework for the last draft of IUCN's *Strategy* and were to a large degree written by Poore.[59]

In the sourcebooks, Poore and his coauthors drew upon a new hypothesis from within the larger field of ecology. This line of argumentation not only pointed at the stability of natural ecosystems, but also tried to explain it. Promoters of this type of research argued for a causality between the biological diversity—meaning the "richness" or miscellaneousness of different biological groups present in an ecosystem on population, species, and gene level—and the stability of natural communities.[60] This reasoning about biological diversity, today known by the shorter phrase *biodiversity,* merged broader ecosystem ecology and narrower research on particular species populations. It also addressed the type of scientific work on genetics and genetic conservation that was monopolized by FAO. It seemed to provide arguments for the ecological value of entire and intact ecosystems and the necessity to protect these and the wildlife they contained—plants and animals—to maintain the stability of the biosphere. The need to protect the planet's biological diversity was central to Poore's ecological reasoning. That evolution had created speciation and diversity on all biological levels, ecosystems, populations, species, and genes, and that this diversity was relatively high in natural ecosystems, was broadly accepted among ecologists. Yet, the diversity-stability hypothesis had gained additional momentum and renewed scientific validity in the 1950s and 1960s, when Evelyn Hutchinson and Robert MacArthur, two promoters of the Modern Synthesis, began to research local interactions of distinct species and links between species and their environment.[61] This type of ecology was developed in the research field of island biogeography, concerned with ecological and evolutionary processes and the relation between species and particular isolated habitats.[62]

Two ideas of the island biogeographers were considered particularly valuable to conservation efforts, and both became part of Poore's argumentative arsenal. The first one was the so-called *equilibrium theory* advanced by the Canadian ecologist MacArthur and the American biogeographer Edward Osborne Wilson. Based on their research on isolated and semi-isolated islands, the two scientists argued that over time ecosystems reached an equilibrium level of species richness. The smaller the area, the less diverse and less stable this equilibrium was. In a famous paper from 1967, McArthur and Wilson proposed that based on the ideas of equilibriums on large isolated islands, one could draw conclusions on the need for large nature reserves that contained biologically diverse, and therefore relative stable, ecosystems.[63] The idea attracted a lot of attention in ecological and conservation cycles and led to more research on the relation between diversity and stability in the following decade.[64]

A second idea formulated by island biogeographers that attracted the interest of nature conservationists at IUCN linked both stability and diversity to the period it took ecosystems to evolve. In this hypothesis, stability and diversity were seen as coming from an increasingly complex organization that ecosystems naturally acquired over time. Therefore, both stability and diversity were higher in natural, nonmodified ecosystems than in those altered through human influences.[65] To "examine the meaning of 'diversity' and 'stability'" in more detail, a short but densely packed symposium was organized at the Brookhaven National Laboratory in Upton, New York, in May 1969.[66] When in 1974 natural scientists met in The Hague for the First International Congress of Ecology, the themes of diversity and stability again played a dominant role. Although for neither "stability" nor "diversity" was there a precise definition, the participating ecologists mostly agreed that both were produced over long periods.[67] In contrast to the systems ecology proposed by Nicholson and others, this second hypothesis thus seemed to suggest that natural ecosystems possessed an inherent advantage over those recently manipulated by humans.

The ideas of island biographers on the benefits of biological diversity provided a welcome tool for conservationists in general, and for Poore in particular, to promote the preservation of natural areas for conservation and research. The first draft of the *World Conservation Strategy* already proclaimed that large natural areas needed to be protected to prevent disastrous and "irreversible consequences" for the planet.[68] In writing the sourcebooks' second drafts, Poore promoted the value of island biogeographical research for nature conservation, drawing on MacArthur and E. O. Wilson's equilibrium theory and similar arguments made by some of their colleagues. Poore particularly took to heart the plea for large reserves for conservation, made by McArthur's

disciple Jared Diamond, and the call to develop a "rational plan for the preservation of diversity" to secure continued evolution, by the American island
biogeographer John Terborgh.[69]

This focus on securing ecosystems' potential to evolve is important. With
this line of argumentation, Poore and his coauthors suggested that without
conservation, the development of natural resources and the improvement of
the human condition would be impossible in the long run. They presented the
planned management and conservation of diverse landscapes as key to the
prevention of natural disasters.[70] Before beginning a land-development project, Poore stressed, it was necessary to determine the degree to which different
ecosystems were suited for particular types of use.[71] Accordingly, the introduction to the second draft of the *Strategy* prescribed the preparation of national
inventories of ecosystems, species, and living and nonliving natural resources
prior to assigning any use to any piece of land.[72] Original ecosystems and their
gene pools had to be made available for research, while modified ecosystems
were to be used to test and exemplify sustainable use. In short, the draft prescribed a global network of research reserves.[73]

Conservationists at IUCN themselves were working on such a network of
reserves. On a global scale, IUCN's International Commission on National
Parks was developing a monitoring system to collect basic data on protected
areas through a network of observers and correspondents.[74] Pointing to these
existing initiatives, the drafts and sourcebooks called for the protection of
"sample ecosystems for genetic diversity" to "ensure species evolution and the
survival of critical habitats, as well as for sites for research and as standards
against which change could be measured."[75] In this respect, IUCN conservationists wanted the *World Conservation Strategy* to serve as a sort of register to
document the state of global conservation priorities, which needed to be kept
up to date.[76]

Advocating a worldwide network of ecosystem reserves to protect the evolutionary potential of the planet's natural resources, Poore and his colleagues
at IUCN argued against alternative approaches to conservation present in the
Ecosystem Conservation Group. First, with the stability-diversity hypothesis,
IUCN conservationists tried to convince UNESCO scientists involved in the
MAB that they needed to protect ecosystem samples from all different biogeographical regions, not just those at the time considered of particular scientific
relevance or cultural value.[77]

In 1973 and 1974, IUCN conservationists Dasmann, Poore, Budowski, and
Carlton Ray had been part of two MAB task forces for the protection of natural
areas and on criteria for selecting biosphere reserves.[78] Despite the seemingly
shared aim of a world network of reserves, by the late 1970s, IUCN elites were
only moderately positive about the MAB project. The biosphere reserves were
certainly innovative, and IUCN conservationists approved of the flexibility,

the range of conservation measures discussed, and of the consideration of local human populations as ecological factors that needed to be included in the equation rather than expelled from the territory. However, Poore in particular pointed out that "flexibility must not be a reason for license," a point he highlighted at a MAB planning meeting on the Mediterranean in 1977.[79] What bothered Poore and others was the apparent arbitrariness with which areas could be defined as biosphere reserves. According to UNESCO's guidelines, the procedure for selecting biosphere reserves remained adaptable and differed "depending on the nature and the purpose of the reserve."[80] The biosphere reserves thus seemed to corrupt Poore's calls for diverse but large and stable ecosystems based on relatively uniform criteria and a relatively even distribution around the globe. At a meeting of the IUCN commission chairpersons in 1978, UNESCO's approach was criticized as a groundless "move away from management by objectives to management by available means."[81] The stability-diversity hypothesis seemed to provide scientific arguments for the relevance of global coverage over flexibility. Based on these arguments, Poore's drafts of the *Strategy* recommended the establishment of a global network of protected ecosystems, as well as the biological diversity they contained. Deriving the most benefits from the special characteristics of the diverse ecosystems would be a first step in maintaining humanity's wellbeing.[82]

A second approach within the Ecosystem Conservation Group that Poore and his coauthors opposed concerned FAO's work on genetic resource conservation ex situ. By the late 1970s, attempts to preserve genetic material in artificial seed banks were well underway. The conservation of plant seeds had become part of modern agriculture in the context of the Green Revolution, first promoted by breeders' organizations from Europe and the United States and later by international organizations like FAO.[83] In 1967, the IBP Section for the Productivity of Terrestrial Communities (IBP/PT) and FAO had held a conjoint technical conference on plant genetic resources for agricultural improvement. Plant geneticist and former IBP/PT convener, the Austrian-Australian Otto Frankel pushed FAO to the front of the new field of plant genetic resource conservation.[84] By the mid-1970s, in addition to FAO, the Consultative Group of International Agricultural Research (CGIAR) also was working on conservation ex situ. CGIAR had been founded in 1971 with strong links to the United Nations Development Program (UNDP) in order to improve the productivity of crops important in the diets of low-income populations in less developed countries.[85] In 1973, the members of CGIAR established an International Board for Plant Genetic Resources.[86] With new, substantial research in plant genetics on the rise, conservationists at IUCN feared that ex situ conservation could undermine their efforts to protect biological diversity in ecosystem reserves on the ground—in situ—especially where land was scarce and economically valuable.

For Poore, but also for others at IUCN, the *Strategy* was the organization's chance to map out the role of ecosystem conservation vis-à-vis genetic conservation in seed banks, while it allowed for emphasis on the need for a global approach that went beyond the conservation of natural resources in developing countries only.[87] The second draft of the *Strategy* therefore put heavy emphasis on the ecological value of a preserved genetic diversity in nature. It proclaimed that "both the maintenance of and improvements in the productivity of intensively farmed resources" depended "in large measure" on "the conservation of uncultivated" species in the ecosystems in which they naturally occurred.[88] Wild species were presented as "essential breeding material" that could be used to ensure that improvements in "yields, nutritional quality, pest and disease resistance, responsiveness to different soils and climates, and other qualities are achieved."[89] Tropical forests, for example, as Allen had explained in the *IUCN Bulletin* already in 1975, were "rich stores of diversity, containing potential foods, drinks, drugs, medicines, gums, resins, specific pesticides and other chemical components," and were better storage options than ex situ seed banks.[90] Therefore, the conservation of a system of ecosystem samples in various stages of modification, Poore's sourcebooks explained, was a prerequisite for the long-term fulfillment of basic human needs. Ecosystems contained the planet's evolutionary potential in the form of natural gene banks and served as a buffer against environmental degradation, droughts, floods, and erosion.[91]

IUCN scientists presented a world network of reserves to protect the world's genetic and biological potential as the means to balance the outer and the inner limits of the human environment. Knowing and maintaining the biological diversity of the planet was presented as the key mechanism to protect basic human needs threatened by natural disasters and inefficient land management. The idea was central to IUCN's second draft of the *World Conservation Strategy*, which, with a few minor changes, was presented to the members of the Ecosystem Conservation Group in 1978. The draft demanded the political commitment to conservation by national governments, which were called upon to devise their own national strategies. Conservation and ecology were to be integrated early in the general education process, and administrators should receive the additional training necessary to compile national inventories of ecosystems, which were supposed to build a foundation for environmental planning. Most importantly, more knowledge about the natural properties and components of ecosystems was needed, and, therefore, research reserves were to be established around the world.[92]

At first, Poore's line of reasoning seemed to be successful. Nature conservationists affiliated with IUCN, but also their colleagues at UNESCO, appreciated the draft, and Poore's call for the conservation of ecosystem samples was equally positively received by some in the periphery of internationally organized conservation. African conservationists such as William Banage—

Ugandan zoologist, professor at Zambia University, and a member of IUCN's Ecology Commission—had repeatedly criticized the seemingly naïve approach of development organizations such as UNEP. Banage felt that these organizations usually did not take into account that environmentally sound development often failed due to complex bureaucratic or political systems within the Southern countries themselves. He believed what was needed was an internationally organized agenda, such as the *World Conservation Strategy*.[93]

With the other member organizations of the Ecosystem Conservation Group, IUCN's draft found a similarly positive resonance. At an IUCN council meeting on 4 May 1978, UNESCO's representative, the ecologist Bernd von Droste, expressed his full approval. In particular, he liked the references to the MAB and identified UNESCO's objectives in strengthening the role played by biosphere reserves in the *World Conservation Strategy*. UNESCO's deputy assistant director general of science Michel Batisse, who effectively headed the MAB, agreed and suggested the third draft should not only focus on the international organization of conservation within the Ecosystem Conservation Group and other institutes as proposed by IUCN, but also use the prospect of additional biosphere reserves to encourage national responsibilities and action.[94] At IUCN's Fourteenth General Assembly and Technical Meeting in September 1978, the drafts were presented to attending members of UNESCO and UNEP. Also here, the feedback was mostly positive. UNEP, represented by former IUCN member Mona Björklund, felt that the drafting was well on its way. Another UNEP representative, Sveneld Evteev, pointed out that the proposed establishment of a network of reserves should be further assisted by UNEP. The only somewhat critical comments related to finding a proper tone and format for the final draft and to arranging a timely launching of the *Strategy*.[95] Based on these "fruitful consultations" with UNEP and UNESCO, IUCN Director General Munro envisioned that within the same year, the final document would appear.[96] After several rounds of minor revisions in 1979, IUCN presented its final text to all members of the Ecosystem Conservation Group.[97]

Sustainable Development through Impact Assessments

When presented to the larger conservation community, the value of biological diversity had generally been appreciated. The text was received differently, however, when passed on to members of UNEP who previously had not been involved in the drafting process.[98] In 1979, Thacher, who in 1975 had replaced Munro as Tolba's deputy at UNEP, commissioned two external reviewers to write a report on IUCN's draft of the *World Conservation Strategy*: Thomas F. Power Jr., UNDP representative in Bangkok, and Jorge Morello, a Brazilian ecologist working for Barbara Ward's International Institute for Environment and Development (IIED). Their report was far less positive than the comments that IUCN scientists had previously received on their *Strategy*. In order to understand the opposition to IUCN's drafts from fractions within UNEP, one has to look at UNEP's scientists' own programs and their position within the UN structure. The critique by UNEP's consultants entailed more than just a disagreement with the conservation principles formulated by Poore and Allen. Instead, several vocal members of UNEP had developed their own scientific program and organizational agenda.

A first major point in IUCN's *Strategy* with which the two consultants found fault was the aspiration for undifferentiated and, to a certain degree, centrally planned conservation schemes, which IUCN conservationists had built into the *Strategy*. Power and Morello particularly targeted those efforts of Poore and Dasmann that had tried to systematize the case-by-case work of the MAB and FAO using one global strategy and proposing a global network of reserves. In particular, the limits and characteristics of biogeographic regions needed refinement, as they were "too broad for effective political action for their protection."[99] While IUCN scientists had been drafting a *Strategy* in a way that portrayed international conservation work as essential to guaranteeing basic human needs, UNEP officials had become increasingly cautious about demands for global strategies in general. UNEP officials had maintained close working relationships with members of UNDP, UNCTAD, and ECOSOC, and, rather than dividing the global biosphere along biogeographical lines, UNEP and its UN partnering organizations continued to be concerned with international negotiations over the rift between developed and less developed economies.

In this respect, the promotion of decentralized planning by some UNEP scientists had developed out of political events in the mid-1970s. In May 1974, the UN General Assembly had published a Declaration on the Establishment of a New International Economic Order and a related Action Program.[100] Under Thacher and Tolba, UNEP, as the environmental consciousness of the UN, was continually involved in discussions about the needed economic reforms. At the time it was not clear, however, what exactly such a new order would mean for international organizations, or how and by whom it should be implemented. In 1976, for instance, the Dutch economist Jan Tinbergen published one interpretation, titled *Reshaping the International Order*. It demanded the establishment of an independent world food authority for the distribution of natural resources and products, and caused a wave of criticism from the newly independent UN member states, which were against any form of centralized authority. Consequently, international development experts within the UN began to distance themselves from the rigidity of international treaties and procedures, instead proposing flexible and adaptable guidelines for economic development.[101] Less rigid and locally designed charters, programs, and declarations, implemented on a regional basis by acts of international organizations, became an important tool for international economic cooperation during the second half of the 1970s.[102]

Within the UN General Assembly at large, the rhetoric began to shift toward flexibility and local solutions. The concept of society-dependent basic human needs was used to relativize the strong claims on compensation made by governments from the global South, which questioned the economic hegemony of the developed North.[103] In the years after Stockholm, UNEP had not only become the UN's flagship bureau in the Southern Hemisphere, but also attracted more and more scientists from developing countries. Thus, perhaps even more than the General Assembly, UNEP had to balance concerns by representatives from developing countries about their jurisdiction over their use of natural resources and their plans for economic development on the one hand, and calls for universal reforms and regulations regarding resource distribution and environmental degradation on the other.[104]

By the late 1970s, UNEP's program looked at environmental problems region by region. While UNEP's governing council had chalked up some cross-regional successes, such as the Mediterranean Action Plan, it had also experienced frictions when it came to the implementation of instructional guidelines to manage the distribution of natural resources between states. At a meeting of the governing council in 1978, for example, a draft document by UNEP on the conservation and use of resources that crossed country borders had been harshly criticized by some of the organization's members.[105] In a period of severe regional draughts, representatives of Ethiopia and Brazil especially had objected to binding allocation schemes for resources that impeded their own access to water and crops.[106] Instead, at UNEP's general council meeting of the same year, there was a strong motion to follow the lines of the General Assembly and their opposition to grand transnational schemes related to natural resources, and to instead focus on resource problems country by country and project by project.[107]

Against this background, it is less surprising that not all of UNEP's representatives were inclined to embrace IUCN's *World Conservation Strategy*, which demanded a global, systematic approach to natural resource development. At least some of UNEP's members feared that the *Strategy* would result in a centralized environmental authority, disrespectful of local economies and regional contexts, or in an exclusion of the "people factor." The discontent with IUCN's drafts expressed by Ossamma M. El-Tayeb, an Egyptian microbiologist from Cairo University, at the time serving as UNEP's senior program officer for genetic resources and biotechnology, spoke to the skepticism held by a larger number of UNEP's members. Expressing postcolonial concerns, El-Tayeb proclaimed in an internal note to his colleagues that it was necessary to eliminate "arguments which designate developing countries as external banks of genetic diversity" at the expense of "people's welfare."[108] Despite all administrative and structural changes, IUCN, with its colonial history and its domination by European and American scientists, bore the mark of an imperial past. According to the two UNEP consultants Power and Morello, both development experts with ample experience in the developing world, most of the professional staff of IUCN lacked a proper understanding of the socioeconomic conditions in the global South, having neither lived there nor worked on local development projects.[109] IUCN's composition, they believed, presented a problem in environmental negotiations between developed and less developed countries.

In trying to build a network different from the Northern-dominated scientific organizations such as IUCN, UNEP had been intensifying its institutional links with SCOPE. SCOPE had emerged in 1969 out of several projects that had begun as part of the IBP. With the Russian soil scientist Victor Kovda as its president and with the Egyptian botanist Kassas fulfilling the role of vice president, SCOPE seemed to offer a new, non-Western approach to international scientific policy advice. The organization's aim was to synthesize information from diverse scientific fields dealing with the environment, and to pursue a number of projects on biochemical cycles, ecotoxicology, and environmental modeling. Scientists involved in SCOPE had begun working with UNEP soon after the latter's establishment in 1972. One of the earliest joint projects of the two organizations was a monitoring mechanism, the Global Environmental Monitoring System (GEMS).[110]

Different from IUCN's global vision, UNEP and SCOPE had focused their cooperation on the particularities of environmental and developmental problems in the global South. In February 1974, UNEP and SCOPE held a Symposium on "Environmental Sciences in Developing Countries" in Kenya, Nairobi.[111] Country-specific environmental problems and development-oriented solution strategies made up the core of their work. Moreover, the symposium was celebrated by its participants for giving a stage to scientists from less developed countries to think "free from the attitudes of the 'rich man's' environmental concepts and precepts, and to create a body and a system of thinking on environmental issues that [would] fit with the conditions prevailing in the poor peoples' environment."[112] With the financial support of UNEP, SCOPE had invested efforts in the development of a methodology that would prevent the breaching of environmental limits and was not based on setting aside natural ecosystems by principle. In fact, since the mid-1970s, SCOPE scientists claimed that the functioning of the biosphere could be protected by assessing and preventing the environmental risk potentially caused by every individual development project ad hoc.[113] When the IUCN's final draft of the *World Conservation Strategy* was presented to UNEP, SCOPE's work on environmental impact assessment presented a viable, alternative take on bridging environmental and developmental concerns to IUCN's suggested world network of reserves for biological diversity.

With SCOPE as a partner, UNEP not only began to work with a scientific community different from IUCN, but also developed a second scientific method to achieve sustainable development. Already Power and Morello had suggested that instead of preserving stability, environmental research was needed to study the "minimal recovery status" of both ecosystems and socioeconomic systems.[114] Starting in the United States in the 1960s, early forms of environmental impact statements had become part of national and international environmental law. However, in the early 1970s, these reports were often conducted only after a project had been executed. Nonetheless, the environmental impact statement requirement under the National Environmental Policy Act of 1969 (NEPA) had brought about widespread interest in methods for assessing the impact of individual development projects on their direct environment.[115] Research related to environmental impact statements moved ahead in disciplines such as biochemistry and biochemical engineering, using new techniques of computer modeling and satellite technology.[116] Members of UNEP attempted to bring environmental impact assessment into the development planning process.[117]

In 1975, UNEP provided the funds for SCOPE scientists to establish a Monitoring and Assessment Research Centre at Chelsea College, London, to test such assessment methods.[118] SCOPE scientists working there defined environmental impact assessment as the "activity designed to identify, interpret and communicate information about the impact of an action, on man's health and well-being (including the well-being of ecosystems on which man's survival depend)."[119] The impact was measured quantitatively as the "net change (good or bad) in man's health and well-being."[120] Drawing on this work by SCOPE, in 1977 the UNEP governing council approved the inclusion of research on the "response of terrestrial ecosystems to pressures exerted on the environment" in the tasks of GEMS, the joint environmental monitoring program run by UNEP and SCOPE scientists.[121] Knowing natural ecosystems' ability to resist a certain amount of pressure was important to the assessment of environmental risk in development projects. Unlike Poore's efforts to secure the biological potential of the biosphere, SCOPE's approach considered the "conflicts between human goals and natural processes" to be largely unavoidable.[122]

SCOPE's approach did not aim at securing biological preconditions for growth, but instead tried to understand how much damage human resource use could do without distorting the balance and the benefits of natural ecosystems. In fact, scientists at SCOPE and UNEP were influenced by scientific research that opposed the diversity-stability hypothesis on which IUCN scientists built their claims. Leading in this criticism was the Canadian ecologist

Crawford "Buzz" Holling. In the early 1970s, Holling published a scientific paper on the stability of ecosystems in which he postulated the idea that ecosystems had not one but several moments of stability and that there was no such thing as an absolute climax, or "benign nature."[123] Instead, he understood ecosystems as being in a state of more or less resistance to collapse, and this was independent of their presumed natural state or diversity. Therefore, it made more sense to speak of "resilient" than of "stable" ecosystems.[124] This, Holling thought, was the case for all ecosystems, natural and modified.

Holling and his ideas on environmental resilience had been first discussed and linked to the method of impact assessment in a SCOPE workshop in 1974.[125] In the following years, UNEP, under Thacher, provided additional funds for Holling and his colleagues at the International Institute for Applied Systems Analysis (IIASA) to develop an environmental impact assessment methodology that could be adapted to different kinds of development projects. In 1978, Holling published a report, titled *Adaptive Environmental Assessment and Management*, which introduced an assessment procedure that could be adapted at the beginning of each development project. The report was based on the findings of a small working group that Holling had called together on UNEP's request. For different case studies, this working group generated indicators relevant for policy decisions, including social, economic, resource, and environmental factors. Based on these factors, alternative development scenarios were developed, best and worst outcomes predicted, and management decisions made.[126]

Environmental impact assessments were not only based on a different idea of environmental stability than that prevalent within IUCN. Holling's approach also emphasized the need for an implementation of his assessment method that could be tailor-made to fit each individual development project. Holling repeatedly pointed out the diverse social, economic, and environmental objectives that might be at stake in development projects in different parts of the world. While his report prescribed a guide for procedure, he stressed that a case-to-case application was needed for effective environmental impact assessment.[127] Pleased with the flexibility entailed in Holling's approach, UNEP scientists began to adapt it to their own environment- and development-related work. In late 1978, members of UNEP initiated a project on guidelines for industry sites in natural areas. At three expert meetings, it was agreed that the guidelines should not be a cookbook-type manual that would recommend a rigid step-by-step approach to environmental impact assessment. As each industrial project would be somewhat different and sited in a different type of environment, a flexible approach, à la Holling, would be needed to carry out assessments of the environmental risk at each proposed site.[128]

In the larger framework of international policymaking, this tailor-made approach was much appreciated. The type of environmental work proposed by UNEP and SCOPE scientists, based on individualized impact assessments, was attractive for countries in the developing South, who were demanding environmental schemes that took into account their difficult economic conditions. In general, governments of the then-called developing countries were skeptical of environmental schemes that demanded a high degree of international cooperation and coordination, such as IUCN's network of reserves for biological diversity.[129] In contrast, UNEP's work on environmental impact assessment promised a type of environmental politics that allowed for local decision-making, rather than binding international agreements. Sustainable development through the assessment of environmental risk in individual development projects as suggested by UNEP did not depend on global schemes or treaties. Moreover, environmental assessment, which allowed for a largely decentralized implementation, could take into account the particular socioeconomic conditions of less developed countries and entrust responsibilities to domestic ministries rather than international organizations. Furthermore, UNEP officials could operationalize this decentralized way of linking environmental limits and basic human needs to strengthen UNEP's cooperation with other members of the UN family by working through their regional committees and offices around the world.[130]

UNEP's focus on environmental impact assessments differed from IUCN's work on the protection of biological diversity in another important way. Questioning the link between diversity and stability, Holling's line of argumentation undermined Poore's plea for the ecological value of diversity in nature. Instead, the work of UNEP and SCOPE scientists on environmental impact assessments assigned a monetary value to separate natural resources, quite opposite to what IUCN conservationists such as Poore had in mind. In 1978, Thacher had asked Holling to find a way to quantify negative impacts on environmental systems, including long-term social and environmental "effects and other deficiencies."[131] Throughout the late 1970s, members of UNEP kept encouraging environmental scientists at workshops and congresses to develop methodologies for impact assessments that would include a thorough quantification of environmental processes' financial cost. Thacher in particular had prompted research into cost-benefit analyses at the Second International Congress on Ecology held in Israel in 1978.[132] In 1980, UNEP's own governing council reported on several other attempts to integrate a cost-benefit analysis into environmental assessment procedures.[133] In January 1980, for example, members of UNEP organized an expert meeting on the "application of cost-benefit analysis to development activities" in New York.[134] Measuring the value of separate categories of natural resources, rather than valuing naturally diverse and stable ecosystems, UNEP's approach put a price tag on environmental degradation or the absence thereof.

The Strategy and Its Legacy

The impact of UNEP's alternative approach can clearly be seen in the final version of the *World Conservation Strategy* that was eventually published in 1980. In many ways, the *Strategy* deviated from IUCN's drafts, which reveals the lasting influence of Power's and Morello's critique. In the final text, the emphasis on biological diversity no longer necessitated the establishment of a global and centralized network to protect samples of the world's ecosystems. This was a significant break, not only with IUCN's drafts and scientific sourcebooks, but also with those earlier calls for global conservation schemes by Nicholson, Dasmann, and Budowski, which had informed Poore's work. The critique by UNEP's consultants on IUCN's draft had also led to the inclusion of a new aim on the first pages of the final *Strategy*, namely the "maintenance of essential ecological processes and life-support systems."[135] This focus on sustaining ecological processes, rather than protecting the diversity of living organisms and habitats, clearly echoed some aspects of Holling's work on resilience. Rather than stressing the link between the diversity of ecosystems, their stability, and therefore the need to protect their long-term potential, this new aim justified a limitation of protection efforts to the immediate usefulness of ecosystems in line with the methods of impact assessments and cost-benefit analyses.

Similarly, the second and third main objective of the *Strategy* also deviated from Poore's original draft. A second aim referred to the protection of genetic diversity and therefore at first sight resembled Poore's demands for preserving the natural diversity of different ecosystem types. No strict distinction was made, however, between ex situ protection in seed banks and the in situ protection of the biological diversity contained in natural ecosystems. The third objective of the *Strategy* concerned the protection of ecosystems and resembled IUCN conservationists' original aims most closely. However, it focused on the sustained utilization of ecosystems rather than on their integration into international conservation schemes.[136] Thus, the use of the term *ecosystem* here seemed to relate as much to UNEP's interpretation of local ecological support systems for a sustainable resource use as it did to Poore's idea of the ecosystem as the gold standard for nature reserves.

While some aspects of Poore's drafts remained in the final *Strategy*, his main concern, namely the protection of ecosystem samples to preserve the natural resource potential of the global environment, was no longer part of the *Strategy*'s main objectives. Instead, the *Strategy* to a significant degree drew on the new approach of tailor-made environmental impact assessments that allowed for regional or local decisions on what kind of nature needed protection. This flexible approach was clearly opposed to conservationist calls for global, standardized networks of protected areas. In this regard, the *Strategy*'s map section

was telling. Unlike earlier IUCN maps on biogeographical regions by Dasmann or Udvardy, it only portrayed those areas that were under immediate
environmental risk.[137]

The *Strategy* singled out particular areas or ecosystem types where transnational cooperation was needed—for instance, in areas around water basins or
transborder forests, and the so-called *global commons*, including the oceans,
the arctic, and the atmosphere. Yet only in these cases were transnational or
international agreements encouraged. The *Strategy* provided little advice on
how global environmental efforts could be further coordinated. Here, too, the
emphasis was on an ad hoc implementation rather than long-term planning.
In general, the *Strategy* mainly appealed to national governments to include
environmental considerations when formulating new development policies.[138]
However, the drafting of binding national strategies modeled after the *World
Conservation Strategy*, as IUCN had envisioned, had been abandoned in the
final text. Instead, the development of such national conservation strategies
depended much on the influence and dedication of national conservation
communities.[139]

The deviation from Poore's drafts did not result in a complete rejection on
the side of IUCN's members, however. In fact, the *Strategy* was formulated
in such broad terms that it could be shared, celebrated, and promoted by all
members of the Ecosystem Conservation Group.[140] The *World Conservation
Strategy* had brought about a common discourse among the members of the
Ecosystem Conservation Group, linking the themes of environmental protection and the guarantee of basic human needs—the outer and the inner
limits—through the concept of sustainable development. The new shared concept, however, did not contribute to a unification of environmental approaches
or implementation strategies between IUCN and UNEP. In fact, the *Strategy*'s
broad scope and vague formulations made it noncommittal. The interpretive
flexibility of the *Strategy* became apparent immediately after its publication.
Within the organizations, different approaches to sustainable development
continued to exist, highlighting contrasting aspects of the *Strategy* in their own
communications. Conservationists with links to IUCN continued to stress the
importance of protecting samples of biological diversity around the world,
whereas UNEP's scientists focused on environmental impact assessments for
local development projects.

The continued dominance of the diversity-stability argument was very apparent, especially in IUCN's internal publications among conservation-minded
scientists. For instance, conservationist Peter Scott from the World Wildlife
Fund (WWF), which had supported the drafting process financially, expressed
his disappointment with the *Strategy*'s pragmatism.[141] Lee Talbot, who had
been in the *Strategy*'s final review committee, continued to repeat IUCN's original rhetoric on a separate ecological value of nature. Writing for the larger

conservation community in 1980, he explained how national parks were to serve as "parachutes" in maintaining sustainable resources for human needs.[142] In IUCN's publications on national parks and nature reserves, the conservation of biological diversity remained an important objective. IUCN's National Parks Commission continued to campaign for a network of substantial, large reserves to capture the whole variation of ecosystems, both in stable climax and modified conditions.[143] At the 1982 International Congress on National Parks in Bali, arguments on the use of parks to preserve the natural potentials of the biosphere were a dominant theme. A telling example was the contribution by Robert and Christine Prescott-Allen on "protected areas as in situ genebanks for the maintenance of wild genetic resources."[144] On top of this, arguments were made for protecting biological diversity as a natural insurance against the degradation of ecosystems by floods or droughts.

Later in 1982, biological diversity protection celebrated a first big success when the UN General Assembly accepted IUCN's World Charter for Nature.[145] This achievement was partially made possible by the rise of the new scientific field of conservation biology, inspired by the same biogeographical approach to biological diversity that Poore had drawn on in the drafting of the *Strategy*. Conservation biology attracted those conservationists, who, like Poore or IUCN's marine ecologist Carleton Ray, demanded more outspoken political engagement by scientists for nature protection.[146] In 1992, new interest and new research in the field of island biogeography resulted in the Global Convention on Biological Diversity, which was opened for signatures at the United Nations Conference on Environment and Development in Rio.[147] What remained central to IUCN's work on biological diversity was the ecological and potential value that it assigned to natural variety, and the need for centralized planning in its conservation. The catchphrase of "keeping the options alive" became central to IUCN's conservation advice.[148] Although ecological arguments on the stability of ecosystems through diversity continues to be contested, the potential benefit of biological diversity for ecosystems and society is still used by IUCN to combat strictly economic and human-centered interpretations of sustainable development. Even today, IUCN promotes the protection of biodiversity for its "intrinsic value" and as a "cornerstone" of human existence and a precondition to achieving the UN's Sustainable Development Goals.[149]

In the years after the publication of the *Strategy*, scientists working for UNEP continued to focus on developing methods for decentralized impact assessments.[150] In 1980, UNEP's Regional Office for Asia and the Pacific, together with the UN Asian and Pacific Development Institute, published a test study with models for environmental assessment statements.[151] This model was taken up in another workshop in November 1982 by the UN Asian and Pacific Development Center in Kuala Lumpur. The result was an assessment form for development projects, very similar to that proposed by Holling in 1978. It included a scoping process on critical variables, a prediction made for various scenarios on resources used, enhanced, or exhausted, and a management plan that was based on a cost-benefit analysis, all with a strong quantitative focus. Different types of ecological processes, such as clean water supply or productive soils, were assigned an assumed monetary value. Wildlife featured in this assessment method, alongside air, water, soil, minerals, forest, and other types of natural resources that were to be ranked according to their economic value. Decisions on which resources to develop and which to protect, which to include in the equation and which to leave out, were supposed to be made by responsible local authorities, based on their own good judgments.[152]

These projects and workshops provided the background material for what was called the Nairobi Declaration, which UNEP released in 1982 on the decennial of the Stockholm Conference. This policy document promoted environmental impact assessments as one of the main means to secure "sustainable development up to the year 2000 and beyond."[153] As a consequence, the process of conducting impact assessments was integrated into the work of many UN agencies dedicated to various aspects of the environment, natural resources, or agriculture. For 1983 and 1984, FAO's Interdepartmental Working Group on Environment and Energy, for instance, initiated an investigation of the environmental impact of the FAO field programs and possible strategies of assessing and mitigating these impacts.[154] A common feature of all these projects was that the final decision-making was done locally and based on a cost-benefit analysis that assigned a monetary value to the environmental processes that were to be developed or preserved.[155] In 1987, then, the same type of environmental assessment became an integral part of the recommendations published in the UN report *Our Common Future*. The report by the World Commission on Environment and Development, also known as the Brundtland Commission, still underlies the UN's stance on sustainable development.[156]

Within UNEP and the UN at large, conducting environmental impact assessments remains a key step in the decision-making process up to the present day, and in many cases an environmental impact statement has become a prerequisite for UN development funding.[157] Environmental impact assessments have, for example, become important for agricultural development programs that involve genetically modified plants.[158] In the late 1990s, UN Secretary General Kofi Annan called for a major study of the human impact on environments around the world. Consequently, the Millennium Ecosystem Assessment (MEA) was launched in 2000. The MAE popularized the term *ecosystem services*, a concept that emphasizes the economic benefits societies gain by keeping individual ecological processes intact.[159] Local environmental impact assessments and the idea of ecosystem services play an important role in the UN's Sustainable Development Goals of 2016, particularly in relation to internationally shared resources and natural disasters.[160] Within the UN, then, environmental impact assessment has become one of the primary means to guarantee sustainable development.

Conclusion

In contrast to what its title suggests, the *World Conservation Strategy* did not bring about the shared approach to global environmental governance for which conservationists affiliated with IUCN, such as Poore, Dasmann, and Budowski, had hoped. The *Strategy* had been intended to devise a way of balancing concerns about the limits to natural resources with concerns about basic human needs in the work of the main international organizations dealing with the natural environment: UNEP, UNESCO, FAO, and IUCN. It is, therefore, not very surprising that much of the existing scholarship has focused on the ways in which the document brought forth the concept of sustainable development, a concept that dominates the discourse on environment and development until the present day. However, an evaluation of the *World Conservation Strategy* in terms of a singular and unifying idea of sustainable development falls short in explaining the discrepancy between the approaches behind the *Strategy*. In contrast, this chapter has proposed an alternative interpretation of the *Strategy* and its legacy in terms of conflicting scientific theories, different scientific networks, and opposing ways of assigning value to natural resources and environmental processes. Focusing on the scientific discourses underlying the different drafts and versions of the *Strategy,* this chapter has shown that the final document in fact manifested two entirely different organizational profiles.

IUCN conservationists surrounding the plant ecologist Poore were trying to integrate their conservation advice with the work of UNESCO scientists on biosphere reserves and the work of agricultural geneticists at FAO. For this, they drew on a new discourse from the scientific field of island biogeography. Based on the argument that natural ecosystems are more diverse and therefore more stable, Poore and his fellow IUCN conservationists assigned a primary ecological value to biological diversity. According to this line of argumentation, protecting samples of biologically diverse ecosystems around the world, including their evolutionary potential, was a prerequisite for sustainable development and the fulfillment of basic human needs.

At the same time, UNEP scientists had developed a second approach of balancing the inner and the outer limits to the environment. In this, UNEP's members were strongly influenced by calls for decentralized and more customized guidelines for environment and development issues that emerged within the UN in the late 1970s and linked to discussions about a New International Economic Order. Instead of global conservation schemes, the members of UNEP proposed environmental impact assessments as a means for integrating environmental factors into the development process. In contrast to IUCN's global network of ecosystem samples and the ecological value assigned to biological diversity in natural systems, UNEP's environmental impact assessments included the integration of environmental processes into the market economy.

While this chapter has shed new light on the different methodologies devised by international environmental institutions at the time of the *World Conservation Strategy*, this analysis has implications that go beyond the historical case itself. It questions the widely shared understanding of sustainable development as a compromise between conservationists and developers. The two different approaches to sustainable development promoted by the members of UNEP and IUCN did not result in an actual compromise, but continued to exist in parallel. In fact, present-day discourses on biodiversity and ecosystem services, two concepts frequently linked to the idea of sustainable development, still contain the same opposite ways of assigning value to nature. And while, as the economic historian Gilbert Rist has claimed, "it is to its ambiguity that the term 'sustainable development' owes its success," so far the concept has not brought about clear-cut solutions to either the social or the natural consequences of environmental degradation.[161]

Notes

1. John McCormick, "The Origins of the World Conservation Strategy," *Environmental Review* 10, no. 3 (1986): 177; Susan Baker, *Sustainable Development* (London: Routledge, 2015), 18.

2. Literature to be mentioned here includes the works of environmental historians Stephen Macekura, Iris Borowy, Bill Adams, Susan Baker, and John McCormick: Stephen Macekura, *Of Limits and Growth: The Rise of Global Sustainable Development in the Twentieth Century* (Cambridge: Cambridge University Press, 2015); Iris Borowy, *Defining Sustainable Development for Our Common Future: A History of the World Commission on Environment and Development (Brundtland Commission)* (London: Routledge, 2013); William M. Adams, *Green Development: Environment and Sustainability in a Developing World* (London: Routledge, 2008); Baker, *Sustainable Development*; McCormick, "The Origins of the World Conservation Strategy."

3. Martin W. Holdgate, *The Green Web: A Union for World Conservation* (Gland: IUCN, 1999), 149.

4. E.g., see Jevgeniy Bluwstein, "From Colonial Fortress to Neoliberal Landscape in Northern Tanzania: A Biopolitical Ecology of Wildlife Conservation," *Journal of Political Ecology* 25, no. 1 (2018); Benjamin Gardner, *Selling the Serengeti: The Cultural Politics of Safari Tourism* (Athens, GA: Georgia University Press, 2016); Cassie M. Hays, "Placing Nature(s) on Safari," *Tourist Studies* 12, no. 3 (2012); Jim Igoe, *The Nature of*

Spectacle: On Images, Money and Conservation Capitalism (Tucson: University of Arizona Press, 2017).

5. E.g., see Thomas Robertson, *The Malthusian Moment: Global Population Growth and the Birth of American Environmentalism* (New Brunswick: Rutgers University Press, 2012); Mathew J. Connelly, *Fatal Misconception: The Struggle to Control World Population* (Cambridge, MA: Harvard University Press, 2008), 239ff.

6. Craig N. Murphy, *The United Nations Development Programme: A Better Way?* (Cambridge: Cambridge University Press, 2006), 165ff. For a history of failed postwar attempts to implement regional development through outsiders, see Daniel Immerwahr, *Thinking Small: The United States and the Lure of Community Development* (Cambridge, MA: Harvard University Press, 2015).

7. FAO, *The State of Food and Agriculture 2000* (Rome: FAO, 2000), 138ff.

8. Giuliano Garavini and Richard R. Nybakken, *After Empires: European Integration, Decolonization, and the Challenge from the Global South 1957–1986* (Oxford: Oxford University Press, 2012), 177.

9. Murphy, *The United Nations Development Programme*, 155; Giuliano Garavini, "Completing Decolonization: The 1973 'Oil Shock' and the Struggle for Economic Rights," *The International History Review* 33, no. 3 (2011): 483.

10. Garavini and Nybakken, *After Empires*, 180; Vijay Prashad, *The Darker Nations: A People's History of the Third World* (New York: New Press, 2008), 188.

11. FAO, *The State of Food and Agriculture 2000*, 139.

12. W. Warren Wagar, *A Short History of the Future* (Chicago: University of Chicago Press, 1989); Michael E. Soulé, "What is Conservation Biology?," *BioScience* 35, no. 11 (1985).

13. Nils Gilman, "The New International Economic Order: A Reintroduction," *Humanity: An International Journal of Human Rights, Humanitarianism, and Development* 6, no. 1 (2015); Marc Frey, Sönke Kunkel, and Corinna R. Unger, eds., *International Organizations and Development, 1945–1990* (London: Palgrave Macmillan, 2014).

14. Ahmend Mahiou, "Declaration on the Establishment of a New International Economic Order," United Nations Audiovisual Library for International Law, 2011, accessed 23 March 2019, http://legal.un.org/avl/pdf/ha/ga_3201/ga_3201_e.pdf; Murphy, *The United Nations Development Programme*, 155; Prashad, *The Darker Nations*, 189.

15. Hans Landsberg, "Reflections on the Stockholm Conference: Stockholm a Success?" 1972, 5, HUL MFS: Box 41, Folder 405.

16. UNEP, *Report of the Governing Council of the United Nations Environment Programme on Its 1st Session, Held at the Palais des Nations, Geneva, from 12 to 22 June 1973* (Geneva: UNEP, 1973), 36.

17. "Excerpt from Mr. Maurice Strong's Introductory Statement to UNEP's Governing Council, 12 June 1973," HUL MFS: Box 33, Folder 334 "III Aspen 'Outer Limits.'"

18. E.g., Raymond F. Dasmann, John P. Milton, and Peter H. Freeman, *Ecological Principles for Economic Development* (New York: John Wiley & Sons, 1973); E. Max Nicholson, *Handbook to the Conservation Section of the International Biological Programme* (London: IBP/CT, 1968).

19. "Excerpt from Mr. Maurice Strong's Introductory Statement to UNEP's Governing Council, 12 June 1973," HUL MFS: Box 33, Folder 334 "III Aspen 'Outer Limits.'"

20. Thomas W. Wilson, "Illustrative World Problems Lacking Adequate Research," to Maurice Strong, 10 July 1973, HUL MFS: Box 33, Folder 334 "III Aspen 'Outer Limits.'"

21. UNESCO, *Soil Map of the World/6, Africa* (Paris: UNESCO, 1973); FAO, *Shifting Cultivation and Soil Conservation in Africa* (Rome: FAO, 1974); Murphy, *The United Nations Development Programme*, 167.

22. Nick Cullather, *The Hungry World: America's Cold War Battle against Poverty in Asia* (Cambridge, MA: Harvard University Press, 2011), 167; Robin Pistorius, *Scientists, Plants and Politics: A History of the Plant Genetic Resources Movement* (Rome: International Plant Genetic Resource Institute, 1997). Other programs that could be counted as pertaining to the "outer limits" were run by the World Health Organization (WHO), the World Meteorological Organization (WMO), and the Inter-Governmental Maritime Consultative Organization (IMCO). These projects focused on ozone depletion, forest decline and acid lakes, and greenhouse gases and global warming. Stephen O. Andersen, K. Madhava Sarma, and Lani Sinclair, *Protecting the Ozone Layer: The United Nations History* (London: Earthscan, 2012), 6; R. Michael M'Gonigle and Mark W. Zacher, *Pollution, Politics, and International Law: Tankers at Sea* (Berkeley: University of California Press, 1981), 48, 74.

23. UNEP and UNCTAD, *The Cocoyoc Symposium on "Pattern of Resource Use, Environment and Development Strategies," Cocoyoc, Mexico, October 8–12, 1974* (Zug: Inter Documentation Co., 1975).

24. Ibid.

25. UNEP, *The Cocoyoc Declaration Adopted by the Participants in the UNEP/UNCTAD Symposium on Patterns of Resource Use, Environment and Development Strategies Held at Cocoyoc, Mexico, from 8 to 12 October 1974* (Nairobi: UNEP, 1974).

26. Gilbert Rist, *The History of Development: From Western Origins to Global Faith* (London: Zed Books, 2014), 162.

27. Peter M. Haas, *Saving the Mediterranean: The Politics of International Environmental Cooperation* (New York: Columbia University Press, 1990), 168.

28. Maurice Strong to Robert O. Anderson, Chairman of the International Institute for Environmental Affairs, 30 August 1973, HUL MFS: Box 33, Folder 334 "III Aspen 'Outer Limits.'"

29. "Excerpt from Mr. Maurice Strong's Introductory Statement to UNEP's Governing Council, 12 June 1973," 15, HUL MFS: Box 33, Folder 334 "III Aspen 'Outer Limits.'"

30. Maurice F. Strong, *Where on Earth Are We Going?* (Toronto: Knopf Canada, 2000), 141; also see Macekura, *Of Limits and Growth*, 129, on the reservations by U.K. and U.S. government officials toward UNEP's role.

31. Dag Hammarskjöld Foundation, *What Now? The 1975 Dag Hammarskjöld Report: Prepared on the Occasion of the Seventh Special Session of the United Nations General Assembly* (Uppsala: Dag Hammarskjöld Foundation, 1975), 114.

32. Tony Loftas, "The UN's Agents of Change," *New Scientist* 66, no. 953 (1975).

33. Maurice F. Strong, "Progress or Catastrophe: Whither Our World?," *Environmental Conservation* 2, no. 2 (1975): 83–84.

34. Ibid., 85.

35. IUCN, "IUCN Prepares World Strategy," *IUCN Bulletin*, n.s., 8, no. 10 (1977): 59.

36. IUCN, "UNEP-IUCN Link in Ecosystem Conservation" *IUCN Bulletin*, n.s., 6, no. 6 (1975): 23.

37. At the time, IUCN was divided into six commissions: the Commission of Ecology, the Commission on National Parks and Protected Areas, the Species Survival Commission,

the Landscape Planning Commission, the Commission on Environmental Law, and the Commission on Education.

38. IUCN, "IUCN's Programme 1976–1978," *IUCN Bulletin*, n.s., 6, no. 11 (1975): 43.

39. John Perry, "RE Provisional List of Honorary Consultants, Duncan Poore," to Peter Scott, 12 December 1975, WL RF: Box 39, Folder "IUCN SSC 1"; Martha Coolidge to Lee Talbot, 29 April 1977, HUA HJC: HUG (FP) 78.16, Box 7 ABC Series, Folder "C Colleagues and Friends, Talbot, Dr Lee M., wife Martha."

40. "Meeting of Commission Chairmen and Deputy Chairmen, Morges, 31 January–2 February 1978," HUA HJC: HUG (FP) 78.20, Box 22, Folder "IUCN-WWF and Related International Conservation 1978-IUCN ADM: Council, Staff and General."

41. Willem H. Van Dobben and Rosemary H. McConnell, eds., *First International Congress of Ecology* (Wageningen: Centre for Agricultural Publication and Documentation, 1974).

42. Pistorius, *Scientists, Plants and Politics*, 53–55.

43. Hans Landsberg, "Reflections on the Stockholm Conference: Stockholm a Success?" 1972, 21–22, HUL MFS: Box 41, Folder 405.

44. Ibid.; also see Cornelius Disco and Eda Kranakis, eds., *Cosmopolitan Commons: Sharing Resources and Risks across Borders* (Cambridge, MA: MIT Press, 2013).

45. Raymond F. Dasmann, "Stockholm, Bucharest, and Rome—What Next?," *IUCN Bulletin*, n.s., 6, no. 1 (1975): 1.

46. Loftas, "The UN's Agents of Change," 608; Martin W. Holdgate et al., *The World Environment 1972–1982: A Report* (Dublin: Tycooly International Pub., 1982), 250.

47. IUCN, ed., *Proceedings of the Thirteenth (Extraordinary) General Assembly, Geneva 1977* (Morges: IUCN, 1977), 13–14.

48. Former IUCN member Martin Holdgate has called it the "night of long knives." Holdgate, *The Green Web*, 124–29.

49. IUCN, ed., *Proceedings of the Fourteenth General Assembly, Ashkhabad 1978* (Morges: IUCN, 1979).

50. Among further involvements, he also helped establish the World Economic Forum in Geneva. In the late 1970s, as chairman of the International Development Research Center of the Canadian government, he was in touch with the international development constituency. Strong, *Where on Earth Are We Going?*, 153–54.

51. Ibid., 145; Loftas, "The UN's Agents of Change"; Jon Tinker, "World Environment: What's Happening at UNEP?," *New Scientist* 66, no. 953 (1975): 600–3.

52. "Minutes of the 24th Meeting of the Executive Committee, Morges, 9–11 February 1977," HUA HJC: HUG (FP) 78.20, Box 20, Folder "IUCN-WWF and Related International Conservation 1977 IUCN ADM: Extraordinary General Assembly, Geneva, April."

53. Lee Talbot, "Interview Questions," email to Simone Schleper on 7 July 2016.

54. Poore in "Meeting of Commission Chairmen and Deputy Chairmen, Morges, 31 January–2 February 1978," HUA HJC: HUG (FP) 78.20, Box 22, Folder "IUCN-WWF and Related International Conservation 1978-IUCN ADM: Council, Staff and General"; IUCN, *Proceedings of the Thirteenth (Extraordinary) General Assembly, Geneva 1977*, 20ff; Duncan Poore, "Presentation of Progress Report," in *Proceedings of the Thirteenth (Extraordinary) General Assembly, Geneva 1977*, ed. IUCN (Morges: IUCN, 1977), 142.

55. Edward Goldsmith and Robert Allen, *Blueprint for Survival* (New York: New American Library, 1972).

56. IUCN and ICNP, *United Nations List of National Parks and Equivalent Reserves* (Morges: IUCN International Commission on National Parks, 1974); Robert R. Miller and IUCN, *Red Data Book* (Morges: IUCN, 1977); "Minutes of IUCN Council Meeting, Morges, 4 May 1978," IUCNL.

57. Poore had been one of Coolidge's preferred candidates for the job (Harold Coolidge to Maurice Strong, 10 March 1977, HUA HJC: HUG (FP) 78.20, Box 20, Folder "IUCN-WWF and Related International Conservation 1977 IUCN ADM: Extraordinary General Assembly, Geneva, April"). However, as Martin Holdgate recalls, "There was no interview: Maurice Strong persuaded the Bureau that they had the right man" (Holdgate, *The Green Web*, 138).

58. IUCN, "First Draft of a World Conservation Strategy," 1978, 5–7, IUCNL; also see Roger Crofts and Des Thompson, "Professor Duncan Poore—Global Conservationist," *Scottish Natural Heritage*, 12 April 2016, accessed 24 March 2019, https://scotlandsna ture.wordpress.com/2016/04/12/professor-duncan-poore-global-conservationist/.

59. The six sourcebooks discussed both different biogeographical regions and different species. One was concerned with marine ecosystems and was written by Allen, drawing on the work of Carleton Ray. Another one by Allen and his wife Christine Prescott-Allen focused on threatened vertebrates, based on the *Red List of Threatened Species*, and one by the botanists Grenville Lucas and Hugh Synge looked at threatened plants. The other three were mainly written by Poore. One concerned ecosystems in the tropics, based on Poore's own work. Another one on drylands was written by Poore together with steppe ecologist Christopher Dunford. A sixth one on ecosystems, also composed by Poore, linked together the first five sourcebooks.

60. E.g., Margaret Dienes and Robert Brown, *Diversity and Stability in Ecological Systems* (Springfield: Clearinghouse for Federal Scientific and Technical Information, 1969), 25.

61. In the 1930s and 1940s, in what is known as the Modern Synthesis, a new agreement on evolutionary processes had formed, reconciling Mendelian genetics with Darwin's gradual evolution by natural selection. Other renowned ecologists and conservation promoters involved were zoologist Ernst Mayr and IUCN founding father Julian Huxley. Julian Huxley, *Evolution: The Modern Synthesis* (New York: Harper & Brothers, 1943); Ernst Mayr, *Systematics and the Origin of Species, from the Viewpoint of a Zoologist* (New York: Columbia University Press, 1942); Mark Vellend, "Conceptual Synthesis in Community Ecology," *The Quarterly Review of Biology* 85, no. 2 (2010): 184.

62. C. Barry Cox and Peter D. Moore, *Biogeography: An Ecological and Evolutionary Approach* (London: John Wiley & Sons, 2010), 10; Sharon E. Kingsland, "Designing Nature Reserves: Adapting Ecology to Real-World Problems," *Endeavour* 26, no. 1 (2002): 10.

63. Mark V. Lomolino, Dov F. Sax, and James H. Brown, *Foundations of Biogeography: Classic Papers with Commentaries* (Chicago: University of Chicago Press, 2004), 2; Robert H. MacArthur and Edward O. Wilson, "An Equilibrium Theory of Insular Zoogeography," *Evolution* (1963); *The Theory of Island Biogeography* (Princeton: Princeton University Press, 1967).

64. John Spicer and Kevin Gaston, *Physiological Diversity: Ecological Implications* (New York: John Wiley & Sons, 2009), 141.

65. Dienes and Brown, *Diversity and Stability in Ecological Systems*, 33.

66. See the preface to Dienes and Brown, *Diversity and Stability in Ecological Systems*.

67. Van Dobben and McConnell, *First International Congress of Ecology*.

68. IUCN, "First Draft of a World Conservation Strategy," 1978, 5–7, IUCNL; IUCN, "Second Draft of a World Conservation Strategy," 1978, 13, IUCNL.

69. E.g., Jared M. Diamond, "The Island Dilemma: Lessons of Modern Biogeographic Studies for the Design of Natural Reserves," *Biological Conservation* 7, no. 2 (1975); MacArthur and Wilson, "An Equilibrium Theory of Insular Zoogeography"; John Terborgh, "Preservation of Natural Diversity: The Problem of Extinction Prone Species," *BioScience* 24, no. 12 (1974): 715.

70. Grenville Lucas and Hugh Synge, "Sourcebook for a World Conservation Strategy: Threatened Higher Plants," 1978, IUCNL; Duncan Poore, "Sourcebook for a World Conservation Strategy: Tropical Rain Forests and Moist Deciduous Forests," 1978, IUCNL; Christopher Dunford and Duncan Poore, "Sourcebook for a World Conservation Strategy: Drylands," 1978, IUCNL; Dasmann, Milton, and Freeman, *Ecological Principles for Economic Development*.

71. Duncan Poore, "Sourcebook for a World Conservation Strategy: Ecosystem Conservation," 1978, 23, IUCNL.

72. IUCN, "Second Draft of a World Conservation Strategy," 1978, xiv, IUCNL.

73. Ibid.

74. Nicholas Polunin and Harold K. Eidsvik, "Ecological Principles for the Establishment and Management of National Parks and Equivalent Reserves," *Environmental Conservation* 6, no. 1 (1979): 25.

75. IUCN, "First Draft of a World Conservation Strategy," 1978, 5–7, IUCNL; IUCN, "Second Draft of a World Conservation Strategy," 1978, 74, IUCNL.

76. An "in-depth knowledge of all the resources of all protected areas," an article by two IUCN conservation experts proclaimed in 1979, was essential to "master-planning and effective management." Polunin and Eidsvik, "Ecological Principles for the Establishment and Management of National Parks and Equivalent Reserves," 26.

77. World Resources Institute et al., eds., *Global Biodiversity Strategy: Guidelines for Action to Save, Study, and Use Earth's Biotic Wealth Sustainably and Equitably* (Washington, DC: World Resource Institute, 1992).

78. UNESCO, *Expert Panel on Project 8: Conservation of Natural Areas and of the Genetic Material They Contain: Final Report* (Paris: UNESCO, 1973); *Task Force On Criteria and Guidelines for the Choice and Establishment of Biosphere Reserves, Organized Jointly by UNESCO and UNEP: Final Report* (Paris: UNESCO, 1974).

79. Duncan Poore, "Meeting on Man and the Biosphere Programme in the Mediterranean Region, Held at Side on the South Coast of Turkey, 5–11 June 1977," *Environmental Conservation* 4, no. 4 (1977): 310.

80. UNESCO, *Task Force on Criteria and Guidelines for the Choice and Establishment of Biosphere Reserves*, 7.

81. "Meeting of Commission Chairmen and Deputy Chairmen, Morges, 31 January–2 February 1978," HUA HJC: HUG (FP) 78.20, Box 22, Folder "IUCN-WWF and Related International Conservation 1978-IUCN ADM: Council, Staff and General."

82. IUCN, "Second Draft of a World Conservation Strategy," 1978, xiii–xix, IUCNL.

83. Pistorius, *Scientists, Plants and Politics*, 18.

84. Ibid., 48–49. Existing institutes working on similar projects included the All-Union Institute for Plant Industry in Leningrad, the Commonwealth Potato Collection at Cambridge, and the Rockefeller Foundation in the United States, but also the Ghana Exploration and Introduction Service established in the late 1960s, and the collections in Latin America, Argentina, Brazil and Mexico, as well as the National Seed Storage Laboratory for Genetic Resources in Japan.

85. Murphy, *The United Nations Development Programme*, 85.

86. Pistorius, *Scientists, Plants and Politics*, 56–57.

87. "Meeting of Commission Chairmen and Deputy Chairmen, Morges, 31 January–2 February 1978," HUA HJC: HUG (FP) 78.20, Box 22, Folder "IUCN-WWF and Related International Conservation 1978-IUCN ADM: Council, Staff and General."

88. IUCN, "Second Draft of a World Conservation Strategy," 1978, 3, IUCNL.

89. Ibid., 5.

90. Robert Allen, "Viewpoint. Interdependence: The Trend We Cannot Buck," *IUCN Bulletin*, n.s., 6, no. 3 (1975): 9–10.

91. Duncan Poore, "Sourcebook for a World Conservation Strategy: Ecosystem Conservation," 1978, 15–19, IUCNL.

92. IUCN, "Second Draft of a World Conservation Strategy," 1978, xiii–xix, IUCNL.

93. UNESCO, MAB, and UNEP, eds., *Ecological Principles for Economic Development: Proceedings of a Seminar and Workshop Held in Blantyre, Malawi, 19–27 May, 1976* (Paris: UNESCO, 1976), 11–18.

94. "Minutes of IUCN Council Meeting, Morges, 4 May 1978," IUCNL.

95. IUCN, *Proceedings of the Fourteenth General Assembly, Ashkhabad 1978*, 95ff., 118, 21.

96. Munro in "Minutes of IUCN Council Meeting, Morges, 4 May 1978," IUCNL.

97. David A. Munro, "Towards a World Strategy of Conservation," *Environmental Conservation* 6, no. 3 (1979): 169–70.

98. Also see Macekura, *Of Limits and Growth*, 241–42.

99. Thomas F. Power and Jorge Morello, "Evaluation of Projects Contracted for Execution to the International Union for Conservation of Nature and Natural Resources as a Supporting Organization," 1979, xiii, HUL PST I: Box 58, Folder 555.

100. "General Assembly Declaration on the Establishment of a New International Economic Order," *American Journal of International Law* 68, no. 4 (1974): 798–801.

101. Jan Tinbergen, Antony J. Dolman, and Jan van Ettinger, *Reshaping the International Order: A Report to the Club of Rome* (New York: Dutton, 1976); Rist, *The History of Development,* 159; Robert L. Rothstein, *Global Bargaining: UNCTAD and the Quest for a New International Economic Order* (Princeton: Princeton University Press, 2015), 178.

102. Mahiou, "Declaration on the Establishment of a New International Economic Order," 2.

103. Marc Nerfin and Fernando H. Cardoso, *Another Development: Approaches and Strategies* (Uppsala: Dag Hammarskjöld Foundation, 1977).

104. Loftas, "The UN's Agents of Change," 608–9.

105. "UNEP Governing Council. Fifth Session, New York, May 9–25, 1977. Agenda Item 6/6," KIT.

106. Arthur H. Westing and UNEP, eds., *Global Resources and International Conflict: Environmental Factors in Strategic Policy and Action* (Oxford: Oxford University Press, 1986), 105.

107. "UNEP Governing Council. Fifth Session, New York, May 9–25, 1977. Agenda Item 6/6," KIT; "UNEP Governing Council. Sixth Session, Nairobi, May 9–25, 1980. Agenda Item 8/2/Add.2," KIT.

108. Ossamma M. El-Tayeb, "Internal Memorandum," to Reuben Olembo, UNEP Senior Programme Officer, 30 March 1979, cited in Holdgate, *The Green Web,* 151; Kent Nnadozie, *African Perspectives on Genetic Resources: A Handbook on Laws, Policies, and Institutions Governing Access and Benefit-Sharing* (Washington, DC: Environmental Law Institute, 2003), 304.

109. Thomas F. Power and Jorge Morello, "Evaluation of Projects Contracted for Execution to the International Union for Conservation of Nature and Natural Resources as a Supporting Organization," 1979, 25, HUL PST I: Box 58, Folder 555.

110. Peter S. Thacher, "What's Happened since Stockholm," *Environmental Science & Technology* 8, no. 3 (1974): 214; Michael D. Gwynne, "The Global Environment Monitoring System (GEMS) of UNEP," *Environmental Conservation* 9, no. 1 (1982): 39; "UNEP Governing Council. Second Session, Nairobi, March 11–22, 1974, Item 8a of the Provisional Agenda," HUL PST II: Box 65, Folder 611.

111. Daniel M. Dworkin and ICSU, *Environmental Sciences in Developing Countries: Summary Reports and Recommendations, SCOPE/UNEP Symposium on Environmental Sciences in Developing Countries, Nairobi, February 11–23, 1974* (Indianapolis: SCOPE Secretariat, 1974).

112. Samir I. Ghabbour, "SCOPE/UNEP Symposium Environmental Sciences in Developing Countries, Held in the Kenyatta Conference Centre, Nairobi, Kenya, 11–23 February 1974," *Environmental Conservation* 1, no. 3 (1974): 236.

113. Dworkin and ICSU, *Environmental Sciences in Developing Countries.*

114. Thomas F. Power and Jorge Morello, "Evaluation of Projects Contracted for Execution to the International Union for Conservation of Nature and Natural Resources as a Supporting Organization," 1979, xiv, HUL PST I: Box 58, Folder 555.

115. Margaret McCall Skutsch and Robin T. Flowerdew, "Measurement Techniques in Environmental Impact Assessment," *Environmental Conservation* 3, no. 3 (1976): 209.

116. Holdgate et al., *The World Environment 1972–1982,* 441; Paul N. Edwards, *A Vast Machine: Computer Models, Climate Data, and the Politics of Global Warming* (Boston: MIT Press, 2010); Clark A. Miller and Paul N. Edwards, *Changing the Atmosphere: Expert Knowledge and Environmental Governance* (Cambridge, MA: MIT Press, 2001).

117. Crawford S. Holling and UNEP, *Adaptive Environmental Assessment and Management* (Laxenburg: International Institute for Applied Systems Analysis, 1978), 16.

118. "UNEP Governing Council. Fifth Session, New York, May 9–25, 1977. Agenda Item 6/6," 2, KIT.

119. Robert E. Munn, ICSU, and SCOPE, *Environmental Impact Assessment: Principles and Procedures* (Toronto: SCOPE, 1975), 12.

120. Ibid., 21–23.

121. "UNEP Governing Council. Fifth Session, New York, May 9–25, 1977. Agenda Item 6/6," KIT.

122. Munn, ICSU, and SCOPE, *Environmental Impact Assessment,* 16.

123. Holling and UNEP, *Adaptive Environmental Assessment and Management,* 9.

124. Crawford S. Holling, "Resilience and Stability of Ecological Systems," *Annual Review of Ecology and Systematics* 4 (1973): 18–19.

125. Dworkin and ICSU, *Environmental Sciences in Developing Countries*.

126. Holling and UNEP, *Adaptive Environmental Assessment and Management*, 14, 48–56.

127. Ibid., 17–18.

128. M. Nay Htun, "Development of UNEP Guidelines for Assessing Industrial Environmental Impact and Environmental Criteria for the Siting of Industry," in *Perspectives on Environmental Impact Assessment*, ed. Brian D. Clark et al. (Dordrecht: Springer Netherlands, 1984), 235.

129. Ernst B. Haas and John G. Ruggie, "What Message in the Medium of Information Systems?," *International Studies Quarterly* 26, no. 2 (1982): 200ff.

130. John Horberry, *Status and Application of Environmental Impact Assessment for Development* (Nairobi: Conservation for Development Centre, 1984), 111.

131. Peter Thacher, "Keynote Speech for Second International Congress of Ecology, Jerusalem, Israel, 10–16 September 1978," HUL PST II: Box 44, Folder 392.

132. Ibid.

133. "UNEP Governing Council. Sixth Session, Nairobi, May 9–25, 1980. Agenda Item 8/2/Add.2," 2, KIT.

134. UNEP, "Special Ad Hoc Experts' Meeting on the Application of Cost-Benefit Analysis to Development Activities; New York, January 28–30, 1980. Background Paper from 22 January," HUL PST II: Box 95, Folder 902 "Application of Cost-Benefit Analysis to Natural Resource Management."

135. IUCN, WWF, and UNEP, *World Conservation Strategy* (Gland: IUCN, 1980), vii.

136. Ibid., vi–vii.

137. Miklos Udvardy, *A Classification of the Biogeographical Provinces of the World* (Morges: IUCN, 1975); Raymond F. Dasmann, *A System for Defining and Classifying Natural Regions for Purposes of Conservation: A Progress Report* (Morges: IUCN, 1973); IUCN, WWF, and UNEP, *World Conservation Strategy*, map section.

138. *World Conservation Strategy*, section 10.

139. The best example here is probably Max Nicholson's contribution to the U.K.'s Conservation Strategy. Brian D. Johnson and WWF-UK, *The Conservation and Development Programme for the UK: A Response to the World Conservation Strategy. An Overview—Resourceful Britain* (London: Kogan Page, 1983). Nicholson became the chairman of the Programme Standing Committee. Critical of the new flexibility introduced in the final *Strategy*, Nicholson's rigidly planned agenda for the U.K. echoed much of his work for the IBP ("Next Steps in World Strategy for Conservation and Development," 1980, 1–3, AL EMN: Box C. 498).

140. "WWF-UK Agreement," 1982, WL RF: Box 56 WWF-UK; Lee M. Talbot, "The World's Conservation Strategy," *Environmental Conservation* 7, no. 4 (1980); also see Michel Batisse, "Of Mammoth and Men: Conservation for Human Survival," *UNESCO Courier* 33, no. 5 (1980).

141. IUCN, "The World Conservation Strategy at a Glance," *IUCN Bulletin*, n.s., 3 (1980): 38–39.

142. Talbot, "The World's Conservation Strategy," 507.

143. IUCN and UNESCO, *The Biosphere Reserve and Its Relationship to Other Protected Areas* (Gland: IUCN, 1979); Duncan Poore et al., *Nature Conservation in Northern and Western Europe* (Gland: IUCN, 1980).

144. Robert Prescott-Allen and Christine Prescott-Allen, "Park Your Genes: Protected Areas as In Situ Genebanks for the Maintenance of Wild Genetic Resources," in *National Parks, Conservation, and Development: The Role of Protected Areas in Sustaining Society. Proceedings of the World Congress on National Parks, Bali, Indonesia, 11–22 October 1982*, ed. Jeffrey A. McNeely and Kenton R. Miller (Washington, DC: Smithsonian Institution Press, 1984), 634ff.

145. UNEP, *World Charter for Nature: United Nations Resolution 37/7 of the General Assembly, 28 October 1982* (Nairobi: UNEP, 1982).

146. Michael E. Soulé and Bruce A. Wilcox, *Conservation Biology: An Evolutionary-Ecological Perspective* (Sunderland: Sinauer Associates, 1980); Carleton Ray, "Promoting 'the Ecology,'" *IUCN Bulletin*, n.s., 13, no. 1/2/3 (1982): 22.

147. Timothy Swanson, *Global Action for Biodiversity: An International Framework for Implementing the Convention on Biological Diversity* (London: Earthscan, 1997).

148. E.g., Walter V. Reid and Kenton Miller, *Keeping Options Alive: The Scientific Basis for Conserving Biodiversity* (Washington, DC: World Resources Institute, 1989).

149. Robert Fischer et al., *Linking Conservation and Poverty Reduction: Landscapes, People and Power* (Oxon: Earthscan, 2008), xi; IUCN, "International Day for Biological Diversity 2015: Biodiversity for Sustainable Development," 22 May 2015, accessed 23 March 2019, https://www.iucn.org/content/international-day-biological-diversity-2015-biodiversity-sustainable-development.

150. Horberry, *Status and Application of Environmental Impact Assessment for Development*, 210–11.

151. Canaganayagam Suriyakumaran, *Environmental Assessment Statements: A Test Model Presentation* (Bangkok: UNEP Regional Office for Asia and the Pacific and United Nations Asian and Pacific Development Institute, 1980).

152. UN, *Environmental Assessment of Development Projects* (Kuala Lumpur: United Nations, Asian and Pacific Development Centre, 1983), 38.

153. UNEP, *Taking a Stand: From Stockholm 1972 to Nairobi 1982. Declarations on the World Environment* (New York: UN, 1982).

154. Horberry, *Status and Application of Environmental Impact Assessment for Development*, 212–13.

155. UN, *Environmental Assessment of Development Projects*, 22; Horberry, *Status and Application of Environmental Impact Assessment for Development*, 21; Borowy, *Defining Sustainable Development for Our Common Future*, 140.

156. World Commission on Environment and Development, *Our Common Future* (Oxford: Oxford University Press, 1987); Borowy, *Defining Sustainable Development for Our Common Future*.

157. Holdgate, *The Green Web*, 440.

158. N. De Sadeleer, *Implementing the Precautionary Principle: Approaches from the Nordic Countries, EU and USA* (London: Earthscan, 2012).

159. José Sarukhán and Joseph Alcamo, *Ecosystems and Human Well-Being: A Framework for Assessment. A Report of the Conceptual Framework Working Group of the Millennium Ecosystem Assessment* (Washington, DC: Island Press, 2003); Joseph Alcamo and Elena M. Bennett, *Ecosystems and Human Well-Being: A Framework for Assessment* (Washington, DC: Island Press, 2003); Sabine Höhler, "Von Biodiversität zu Biodiver-

sifizierung: Eine Neue Ökonomie der Natur?," *Berichte zur Wissenschaftsgeschichte* 37, no. 1 (2014); Eric Vanhaute, "Van Malthus tot Rio: retoriek rond economie en ecologie," *Jaarboek voor Ecologische Geschiedenis* 1 (1999).

160. UN, "Transforming Our World: The 2030 Agenda for Sustainable Development," Sustainable Development Goals Knowledge Platform, n.d., accessed 21 March 2019, https://sustainabledevelopment.un.org/post2015/transformingourworld.

161. Rist, *The History of Development*, 192–94.

IUCN and Environmental Expertise, 1960s–Present

Following a group of conservation-minded biologists through the changing context of international policymaking from the late 1950s to the mid-1980s, this book has examined the development of their scientific arguments, their proposals for global conservation schemes, and the negotiations of their expert roles. At the time, these conservationists were mainly organized in the International Union for Conservation of Nature and Natural Resources (IUCN) (figure 5.1). This book has studied the negotiations between leading IUCN conservationists, groups of scientists, and policymakers over the role of experts during what has come to be called the *environmental age*.[1] During these decisive decades for international environmental policymaking, pleas to devise international regulations to protect the global environment were increasingly prominent in the international fora of research and diplomacy. Examining these new demands for environmental expertise in international organizations, this book has unraveled the interlacing and negotiation of scientific approaches to nature protection and their politics.

The questions that this book set out to address pertained on the one hand to the scientific work done by IUCN conservationists, including both the negotiations of the scientific foundations of conservation expertise and the strategies for implementing nature conservation in the international sphere, and on the other hand to the sociopolitical roles of nature protection experts in dealing with both social and natural aspects of environmental degradation. The narrative followed a small elite of IUCN scientists through three controversies on conservation science and expert politics: the disagreement among IUCN conservationists about the scientific approach to conservation during the International Biological Program (IBP), which took place between 1964 and 1974; the conflict between different claims to environmental expertise and expert roles by IUCN conservationists and policymakers in early international environmental politics around the time of the United Nations Stockholm Conference on the Human Environment of 1972; and the opposing practices of assigning societal value to nature pursued by scientists in IUCN and the UN Environmental Program (UNEP) during the years leading up to the publication of the

Figure 5.1. *From left to right:* Ray Dasmann, Gerardo Budowski, Kenton Miller, David Munro, Duncan Poore, and Lee Talbot, 1987 (photo ID 2053) © IUCN.

World Conservation Strategy of 1980 and beyond. In doing so, this book has charted how international scientific projects strengthened the understanding of global environmental problems that could be studied by science, while postwar politics of decolonization and the Cold War problematized global solutions and IUCN's ideas on all-encompassing universal expert roles.

It is not a straightforward success story, then, that this book has presented. Instead, it is a story with shades of grey, some successes, and some failures.[2] There have been many recorded achievements in the organization's seventy-year history. Over the years, IUCN has grown to become the largest international nature conservation organization, and its members' efforts have undoubtedly had a significant impact on keeping conservation on the agenda of international environmental politics.[3] The *IUCN Red List of Threatened Species* continues to be the main signpost for biodiversity loss. The organization cooperates with political institutions around the globe, and continues to be a major contributor to international environmental policymaking with a strong focus on biodiversity protection for sustainable development.[4] However, the previous chapters have demonstrated that in intergovernmental settings, IUCN conservationists were at times bypassed in their claims to broad environmental authority, as well as in their attempts to integrate global nature protection schemes into policies on the use of natural resources and their development.

The resulting discontent is trenchantly captured in an editorial that IUCN's senior ecologist Ray Dasmann wrote in 1975 for the *IUCN Bulletin*: "Because of international rivalries and in the interest of short term political gains," Dasmann lamented, the leaders of the nations of the world had "refused to address themselves to the real issue." He added that it was "difficult to be optimistic about such concerns as the conservation of wildlife and natural areas" when political leaders showed such little attention to IUCN's message on the ecological interrelatedness of nature protection and society's wellbeing—the "real concern for the future of mankind."[5]

The mixed success of IUCN conservationists did not lie solely in their ignorance of the United Nations' (UN's) development politics, a factor that has been pointed to by previous scholarship.[6] Throughout the period studied, nature conservationists at IUCN adapted their approaches to protect global nature to broader and socially relevant themes in the political discourses at intergovernmental organizations. These themes included the use of natural resources, debates on technical assistance as part of development aid, and the guarantee of basic human needs. Recognizing the significance of international foreign aid reforms, IUCN members reframed conservation as the ecological precondition for successful economic development. IUCN was concerned with development from an ecological perspective; however, this was at odds with other definitions of societal progress as technical and economic advancement. Main causes for the limited role that IUCN experts were granted within intergovernmental organizations were instead their broad claims to environmental expertise, the emergence of alternative interpretations of environmental systems that favored locally implemented solutions, and conservationists' unreflective assertions of their politically neutral scientific authority. Despite their argumentative versatility, IUCN members lagged behind in questions related to how scientific advice should be implemented and by whom. In fact, until the late 1980s, IUCN remained an advisory organization with few contacts on the ground. At the time, however, centralized forms of environmental governance were increasingly criticized in international fora linked to the UN. IUCN conservationists nevertheless continued to promote top-down approaches to conservation in which international scientific bodies were to be given authority over national policymakers. Insisting on the all-encompassing relevance of ecosystem ecology, their approach to conservation and environmental expertise bore the mark of centralized technocratic elitism, irreconcilable with postcolonial reform politics and a growing international recognition of socioeconomic inequalities.

So far, the relevance of the global and the local have often been stressed separately within large narratives in environmental history, or in studies on grassroots environmentalism.[7] Notions of different scales in environmental thinking, although often playing an indirect role, have found some concrete conceptual discussion in the postmodern cultural critique by authors such as literary scholar Ursula Heise and most recently Bruno Latour.[8] Yet, while global visions are usually dated back to the arising environmental movement of the 1960s, the emphasis on the local is often seen as preceding a planetary understanding of the environment or as forming a reaction to it. In the context of international climate governance, Marybeth Long Martello and Sheila Jasanoff, for instance, have pointed to the "rediscovery of the local" by the scientific and political bodies of the 1990s.[9] However, by studying the history of global nature conservation as promoted by IUCN scientists and administrators, this book has shown that tensions between local and global ideas on governance and expertise were at the core of postwar international environmental politics from their initiation. The tension between a new planetary consciousness and the perceived need for flexible and locally implemented strategies was central to the controversies around conservation advice and environmental expertise that this book has been concerned with.

Already during the IBP, tensions between universalists and localists within the international conservation community became apparent. During the early 1970s, a group of IUCN conservationists, including Raymond Dasmann and Duncan Poore, tried to link claims about the global need for, and the universal applicability of, conservation to a more inclusive politics from below. They did so by promoting the integration of local ecological knowledge and priorities in conservation tools such as local resource maps. IUCN conservationists such as Dasmann, who had studied traditional forms of nature conservation abroad, resented the IBP convener Max Nicholson's focus on the functions and processes of natural systems. Instead, they insisted on the need to focus on the protection of threatened wildlife in particular local topographies as the main responsibility of conservation.

The intra-organizational negotiations between representatives of both traditions led to a new approach to ecosystem conservation in the early 1970s. This blended approach allowed for both species conservation and ecological research in what were called biogeographical regions, which contained natural ecosystems of similar types and species composition.[10] Biogeographical regions presented a strong unifying paradigm for different types of conservation interests within IUCN. It integrated localists' demands for a stronger differentiation between ecosystem types and the diverse conservation needs in different world regions. Yet it also allowed for a global program that covered all world regions, an approach that IUCN continued to pursue from the early 1970s onward. Based on this first larger project of interinstitutional cooperation, in the early 1970s, IUCN President Harold Coolidge and Director General Gerardo Budowski hoped to integrate a global conservation program into the program of the first UN conference dedicated to environmental problems, the UN Conference on the Human Environment. At this landmark conference, held in Stockholm in the summer of 1972, diplomats and policymakers discussed how environmental issues should be dealt with by individual governments and among the UN's member states. In 1972, in the rooms of the Folkets Hus in Stockholm and to an audience of political diplomats, Coolidge and Budowski promoted the relevance of global conservation schemes for successful land and resource development in different parts of the world, based on their ecological knowledge of ecosystems in the different biogeographical regions. Playing according to these ecological rules of the game would prevent breaching the biosphere's biological limits through mismanagement or unreflective development.[11] This line of argumentation connected with the predictions made by other groups of environmentally concerned scientists, such as those grouped in the Club of Rome, on the physical limits of the biosphere that would eventually constrain economic growth and development.[12] At the same time, IUCN conservationists could use their idea of ecological rules for development to address techno-optimists and socioeconomic planners, such as the advisers to the Stockholm Conference, John McHale and René Dubos, who promoted environmental solutions based on the improvement of resource distribution and a more efficient resource use.[13]

However, not all differences regarding the interests and emphases of intellectuals at the Stockholm Conference could be overcome by a flexible terminology. UN environmental policymaking was defined in socioeconomic terms and assigned to separate expert groups in organizations including the United Nations Educational, Scientific and Cultural Organization (UNESCO), the UN Food and Agricultural Organization (FAO), the International Institute for Environment and Development (IIED), and the Scientific Committee on Problems of the Environment (SCOPE).[14] In the resulting distribution of expert responsibilities, conservation was confined to traditional topics, such as the preservation of threatened species living in species-rich and relatively undeveloped regions in the global South, which meant a setback for the globally coordinated conservation work IUCN's elite had in mind.

During the second half of the 1970s, the members of IUCN adapted their ecosystem approach to the demand for specialization and the renewed focus on wildlife, while nevertheless promoting a global scheme for ecologically based resource management. When drafting the *World Conservation Strategy,* IUCN's scientific director Poore presented ecosystem conservation as a means to protect natural wildlife, rather than as an end in itself. Nevertheless, Poore also continued to link his version of ecosystem thinking to the Second United Nations Development Decade by emphasizing the relevance of biological diversity and wildlife for the stability of the biosphere at large. The protection of diverse and stable ecosystems around the world, then, was presented as necessary to safeguard the planet's evolutionary potential. With this approach, framed as *sustainable development,* conservationists hoped to expand the niche to which they had been confined after the Stockholm Conference. However, while a new and versatile terminology of sustainability was born, this did not overcome the fundamental differences in scale when assigning societal value to intact natural ecosystems that underlay IUCN's and UNEP's approaches to natural resources. Until the present day, this difference remains prevalent in discussions on the general value of biodiversity by IUCN conservationists on the one hand, and the tailor-made and local assessment of the economic value of ecosystem services by UNEP scientists on the other.[15]

This tension between universalist and localist approaches that dominated the early years of international environmental policymaking had critical consequences for the designation and design of environmental expert roles. We see that conservation elites who had begun their conservation careers in the 1940s, 1950s, and 1960s saw IUCN primarily as a body of scientific advisers. As discussed in the previous chapters, systems thinking and the importance of protecting the natural environment for the wellbeing of human societies shaped IUCN conservationists' arguments regarding the global relevance of their work. This line of argumentation also determined the roles and responsibilities that conservationists saw for themselves within international policymaking. A strong belief in the political neutrality of their ecological knowledge was at the core of their self-fashioning as universal environmental experts.[16] Throughout the period, conservationists at IUCN presented themselves as unbiased environmental authorities, qualified to fill high-ranking advisory functions in intergovernmental organizations because of their ecological understanding, fieldwork experience, and numerous personal contacts and institutional networks. While in some cases the members of IUCN indeed managed to circumvent political taboos, overall their proclaimed neutrality came to be contested.

Scientists at IUCN have addressed and published for different kinds of audiences, including the concerned public, development planners, policymakers, and politicians. In this, they continued to rely on their existing, broad, but rather homogenous, network of conservation experts. The idea of a science-based, universal, and apolitical conservation regime was perhaps most strongly expressed in the work of Nicholson and his IBP section. Yet all of the conservationists treated in this book insisted on their neutral advisory role as ecosystem experts, avoiding association with any particular political camp or new environmental activist groups. Their own organization remained hierarchical, based on scientific merit rather than on national or regional representation.

In some cases, IUCN conservationists, with their extensive network and contacts in the international scientific community, were indeed more successful in establishing global support for their environmental cause than were organizations based on government membership. As members of a large-scale scientific nongovernmental organization, they could successfully circumvent the Cold War radio silence between East and West in some particular instances. During the IBP, there was scientific cooperation with Soviet scientists, such as Victor Kovda, whom Nicholson could reach through his links with Polish scientists.[17] In fact, many of the founding members of IUCN who still had influence on international conservation discourse in the 1960s and 1970s, including Nicholson and Coolidge, were part of a generation of internationally minded scientists who have been called "technocrat internationalists."[18] As such, they were sympathetic to centralized planning ideas of the interwar period, later associated with Soviet ideology. Yet not only conservationists of IUCN but also officials at UNESCO and the U.S. Department of the Interior saw in IUCN a way to find a loophole for environmental questions beyond the Iron Curtain.[19]

However, international nature conservation in the period studied was much less troubled by East-West relations than by the growing North-South divide, because of the links made between environmental topics and questions on development aid and technical assistance. It was precisely in bridging this second geographical divide that conservationists in IUCN depended on the UN's agencies, such as UNESCO, FAO, and later UNEP. Not only did these agencies possess the funds that IUCN needed, but they also provided the international reach and the contacts with government officials in the Southern Hemisphere that IUCN lacked. While insisting on the political inclusiveness of their organization, the conservation network was in fact hardly grounded in the global South. The work of colonial conservation regimes had often involved the rigid protection of large conservation areas from local human communities.[20] After the decolonization, conservation scientists rapidly lost influence.[21] IUCN at the time included conservationists from South America and Asia, but often these had been trained in the universities of Europe or North America. Despite some individual contacts, even the African continent—which included many of the traditional geographical focus areas for nature conservation, with its large "charismatic" species and migrating herds—was represented by European expatriates rather than native experts.[22]

For IUCN elites, the composition of their network did not present a problem for their universal aspirations. In their way of thinking, all human societies were part of, and dependent on, the biosphere in the same way. Power politics of any kind made little sense to them. In fact, even after increased attempts to liaise with intergovernmental organizations, Director General David Munro announced in a leading article of the *IUCN Bulletin* in 1979 that "IUCN is and must continue to be non-political—in the sense that it cannot have ties with national or international political blocs and its position must be based firmly on the logic of conservation."[23] In the biosphere, the division of Earth's surface was to proceed along the lines of biogeographical regions, rather than according to the borders of nation states. IUCN conservationists, with their knowledge of ecosystem ecology, did not have to be locals to know how to solve conservation problems in geographical regions far removed from the IUCN headquarters in Morges, Switzerland.

In practice, however, while IUCN conservationists successfully liaised with like-minded biologists, often the blank political canvas that IUCN held up was filled in by others' perceptions. In the late 1970s, scientists from the global South and development experts described IUCN's network as elitist and characterized the Union's work as the "promotion of conservation for conservation's sake," irrespective of local socioeconomic realities.[24] In the course of the period studied, the focus of IUCN conservation elites on their global relevance and political neutrality became a problem when they did not see the parameters of scientific expertise around them changing. In this regard, the Western composition of IUCN's network was increasingly seen as controversial when, within the UN, a new political majority was composed of recently decolonized countries. These had mostly little interest in scientific networks like IUCN's, perceiving the Union as old-fashioned and representing mainly Western and Northern concerns. In the second half of the 1970s, UNEP in particular became an environmental organization focused on regional development needs, building a separate network with a strong representation of experts from the global South.[25]

The study of IUCN's history during the environmental age, then, presents us with a picture in which diverse and at times conflicting ideas about the nature of conservation work continue to coexist and challenge one another underneath a shared umbrella of large, yet abstract, environmental concepts. Historians such as Stephen Macekura and Iris Borowy have already pointed to the ambivalent success of the concept of sustainable development, a powerful linguistic device, which, however, has done little so far to change existing policies.[26] In intergovernmental discussions on the limits to growth, there was broader acceptance of interpretations of environmental problems that defined these issues in social and economic rather than ecological terms, thus taking on board the interests of newly independent and developing countries. These development agendas pushed for a fragmentation of environmental concerns into piecemeal factors to be included in local development plans, such as individual ecosystem services. At the same time, tension grew between technocratic ideas on nature conservation, and decentralized, local approaches that allowed for geographical representation in environmental governance from below. The two decades studied mark a period when environmental expertise became subject to international diplomacy. In this respect, scientific neutrality became politically controversial when conservationists at IUCN built on it their claims for universal environmental authority. Notions of expertise as promoted by IUCN elites, which based their justification on notions of universally valid and globally relevant ecological knowledge, carried the mark of colonial networks and insufficient integration, and were thus dismissed. These areas of tension characterized nature conservation during the formative years of international environmental policymaking.

Where, then, does this leave us? By looking at the history of IUCN and environmental expertise during the 1960s and 1970s, we have seen that the rise of a new planetary environmental awareness among public intellectuals of the global North and the calls for global conservation schemes by groups such as IUCN did not to bring about a shared "green" perspective to socioeconomic or political topics. In this respect, this book cuts across notions of a radical ecological turn in postwar international politics.[27] More than that, a look at the work of IUCN and at discussions about environmental expertise within the fora of the UN reveals the origins of the discrepancies between universalist, planetary aspirations and local forms of governance and expertise, which, from the 1980s onward, have continued to trouble environmental and conservation policymaking.

With the growing dominance of tailor-made over one-size-fits-all approaches to global conservation schemes, after 1980, there has not been an internationally accepted strategy for the environment in its entirety. After the limited success of the *World Conservation Strategy*, IUCN followed a strategy of adaptation. In fact, since 1980, IUCN copied UNEP's approach of regionalization, putting in place regional offices and units. IUCN's program during the 1980s and 1990s was still dominated by the implementation of the *World Conservation Strategy*. Yet, IUCN and its members realized that without regional offices, they would not succeed in gaining the support of the UN and its diverse member states. In 1981, IUCN set up a Conservation for Development Centre (CDC) under the British engineer Michael Cockerell, which was to assist with local and community-based development programs in the global South, particularly on the African continent. This new regional unit was both a concession to the demands by UNEP's members, who had criticized IUCN's lack of practical experience, and a significant turn to traditional conservation interests in African wildlife. While there was much support for the CDC among IUCN's broad membership, and the expectation that the center would help channel IUCN's advisory expertise into implementation plans for National Conservation Strategies (NCSs), a fraction of members was disappointed by the emphasis that Strong and Cockerell put on project management rather than on ecological expertise.

This new focus also affected decisions on leadership. During the presidential elections of 1984, the previously favored IUCN vice-president, Abulbar a Gain, a leading systems thinker and the former head of the Monitoring and Environmental Protection Agency of Saudi Arabia, lost the vote to Mankombu Swaminatha, an Indian plant geneticist with close links to UNEP.[28] In the course of the 1980s, while IUCN gained more foothold in the global South through this regional presence, the previous vision for a global approach to ecosystem conservation eroded further.[29] By 1986, it was felt within IUCN that most of the forty NCSs and subnational conservation strategies underway did not sufficiently deal with the topic of saving the world's biological diversity and its value for sustainable development. According to an internal reviewing committee of 1986, the CDC was "too people focused" and "insufficiently" included wildlife and biological diversity into their work.[30] At the same time, most NCSs composed under the CDC were written with only few links to IUCN.[31] Stronger international guidance by means of general handbooks was required, the committee suggested.

By 1990, coauthor of the first *World Conservation Strategy*, Robert Prescott-Allen, was working on a second global conservation strategy with the goal of bringing international coordination back into the regional rhetoric. IUCN's secretary general Jeffrey McNeely and the previous director general Kenton Miller proclaimed that this new "global strategy" for conserving "biological diversity" was essential to integrating "local and regional efforts" while providing "concise guidance" and "global goals."[32] Compared to the first *World Conservation Strategy*, this second strategy focused more strongly on the direct economic benefits of biological diversity. Three years earlier, in 1987, the World Commission on Environment and Development, under the chairpersonship of Gro Harlem Brudtland, had made "sustainable development" a household term.[33] The link to biological diversity, however, was mostly lost, and the main focus was local economic development within limits. The new strategy, called *Caring for the Earth*, to be launched in 1991 in the run-up to the planned UN Conference on Environment and Development (UNCED), thus presented another of IUCN's attempts to reconcile an emphasis on human development in the global South with a global framework.[34]

At the June 1992 UNCED, more commonly known as the Rio Earth Summit, conservationists associated with IUCN had high hopes of achieving international recognition that biological diversity underlay sustainable development. Even though the Brundtland Commission had not brought about the global agreement that IUCN had hoped for, 1987 had nevertheless produced a first step in the direction of international environmental policymaking. With the Montreal Protocol, signed in 1987, all industrialized countries had agreed to avoid or minimize the use of ozone-depleting chemicals. The preconditions for a similar "truly global environment protection agreement" to be developed in Rio seemed promising.[35] In the limelight of the Rio Summit, especially the European Community and Japan showed themselves willing to agree on a clear timeline for a biological diversity program. As at Stockholm, IUCN members were involved early on in both the preparation of the official documents and the planning for the Global Forum for participating NGOs. The Global Forum was much bigger and less ad hoc compared to the Environmental Forum that had accompanied the Stockholm Conference. Again, representatives of IUCN insisted on competence-driven authority for scientific NGOs, but this was still met with only partial understanding by governments.

For a week, the environment gained all the attention, but little came from it. The Convention on Biological Diversity that was issued at the Ro Summit, echoing much of Poore's and others' preparatory work for the *World Conservation Strategy* on the links between biological diversity and ecological stability, was observed with much disappointment in the conservation community, as again no binding agreement was reached by the participating governments.[36] Again, developments in world politics and international relations dominated the conference: this time the end of the Cold War and the consequential re-ordering and Americanization of international policymaking influenced the decisions about environmental measures and expertise at Rio. While in Stockholm the focus on the development of the Third World had disrupted IUCN's global ecological plans, in Rio the politics of the Washington Consensus dominated the debate. No longer was the support for regional development the focus of environmental policy, but the equal integration of less developed countries into a market-based system that rejected all forms of strict regulation. In fact, the U.S. government was one of the biggest hurdles for the conference's success, as legislators objected to any possible binding agreement at or after the conference that would support the demands by newly independent countries for financial reimbursement and technical assistance to integrate the environment into their plans for development. The U.S. government under George H. W. Bush refused any commitment to recompensation schemes, but argued for low trade barriers and open markets as sufficient incentives.

Conservationists' scientific contribution helped little to ease the tension.[37] The aspirations by conservationists to promote environmental protection as the necessary precondition for growth were shattered by this new emphasis on growth as the prerequisite for environmental protection. In general, the market-based approach presented a turn in rhetoric. Previously, the international community had called for more adapted technological aid for countries located in the global South; now the end of the Cold War seemed to justify the United States' call for equal treatment and noninterference. The end result was the same, however. No binding international rules were established at the Rio Conference. In 1994, IUCN published a guide on how to interpret the Convention on Biological Diversity. Participating IUCN members regretted that their interpretation of biodiversity as the "'common heritage' of humankind was rejected at an early stage," when efforts "geared at establishing international mechanisms" were "met with considerable resistance" by the United States, while "the Group of 77" feared that "developed countries" would want to "influence or even dictate action on biological resources under their [national] jurisdiction."[38]

From the 1980s onward, then, the global aspirations that IUCN had represented throughout the period studied in this book continued to be challenged
by a growing politics of market integration and deregulation. But what did this
continued contestation of global conservation schemes mean for the roles that
conservation experts were to play in international environmental politics? In
the case of IUCN, universal schemes for global nature protection had gone
hand in hand with technocratic aspirations, which came to be criticized in the
course of the 1970s, 1980s, and 1990s. A decentralized and regionally focused
framework for conservation expertise, as demanded by UNEP, seemed to allow for the inclusion of different local forms of expertise. And indeed, since the
1980s, notions of expertise within IUCN seems to have become defined more
broadly and inclusively, moving away from Western dominance in ideas on
nature conservation. Fikret Berkes, an ecologist and member of IUCN's Commission on Environmental, Economic and Social Policy, has proposed that
community-based conservation projects as pursued since the 1980s by leading organizations including IUCN, comprising community forestry, fishery,
and local conservation management, have in fact challenged "enlightenment
assumptions of predictability and control" of postwar positivism.[39] While the
relationship between IUCN and indigenous groups has a contentious history,[40]
since the mid-1990s, IUCN has included "cultural resources" in their definition of protected areas and has incorporated the socioeconomic and cultural
values and rights of local communities into conservation measures.[41] Today,
IUCN has adapted the UN's policy on indigenous groups and presents itself
as an open platform where "governments, NGOs, scientists, businesses, local
communities and indigenous people can work together to forge and implement solutions."[42]

While these examples seem to signify a shift away from traditional, technocratic notions of expertise, whether the changed discourse about scientific and
technical assistance from the 1970s has brought about truly inclusive forms
of environmental structures remains a point of discussion. Critics of the UN's
2030 Agenda for Sustainable Development have pointed out that nonbinding
programs put most emphasis on the private sector and that the aim to include
local people as "stakeholders" does not challenge the neoliberal framework that
has created and keeps in place inequalities in power, voice, and wealth.[43] Environmental historians have long highlighted the persisting colonial structures
in the nature conservation business.[44] Forms of local conservation and development projects, often clustered under the loose concept of community-based
conservation, have continued to work within the institutional frameworks that
have previously been "geared towards fortress conservation models," dating
from colonial times.[45]

Alongside the persistence of old networks and the creation of new capitalist elites within global nature conservation, media coverage on indigenous

spokesgroups camouflages that, in general, linear models of science and expertise continue to dominate the self-fashioning of environmental advisors in, but also beyond, the conservation community, in which scientists tend to oversee the values and social preferences embedded in their work.[46] It is true that since 2009, the UN's Intergovernmental Panel on Climate Change (IPCC) has been calling for more open and inclusive policy debates and has organized several events meant to bring together scientists and different societal stakeholders. In 2014, for instance, the IPCC's Future Earth Forum convened scientists from different disciplines and decision-makers from UN bodies, NGOs, and businesses.[47] While in recent years we have thus seen an opening up of discussions about expertise, this does not mean that we have arrived at a truly inclusive environmental debate. Attacks on science by climate change deniers—perhaps unsurprisingly—seem to have encouraged dogmatic calls for scientific neutrality and apolitical authority, with a limited notion of the scientific and environmental expertise, firmly in the hands of the natural sciences.[48] During the UN Climate Change Conference held in Copenhagen in 2009, for example, different interpretations of climate phenomena observed in the Arctic resulted in a vocal clash between members of the Western-dominated IPCC and representatives of the Inuit community. After describing how the sun was now seemingly setting in a different place, suggesting as an explanation that the axis of the Earth might have shifted, local Arctic communities were cautiously kept at a distance from the climate cause.[49] During the spring of 2017, one could observe another example of this renewed scientific "credibility contest"[50] in many university towns, when environmental researchers in different parts of the world left their ivory towers and took to the streets to march for science and against the significant cuts in national funding and the outspoken distrust experienced by the natural sciences under the Trump administration.[51] Unfortunately, the leading rhetoric underlying these marches displayed little interest or support for researchers outside of the hard sciences. Put in the crosshairs by climate change deniers, many marchers retreated to a technocratic presentation of expertise firmly lying in the hands of natural scientists.[52]

It thus seems as if ideas on scientific neutrality and the hegemony of the natural sciences prevail not only within the conservation community but also within the environmental science sector at large, while global nature protection schemes remain politically nonviable. In the early 2000s, shortly before his death, Max Nicholson wrote resignedly on the last pages of his unpublished autobiography, "While thanks to the ecologists and conservationists, environment had ceased to be a strange concept around forty years ago, human use of natural resources still has to reach the stage of familiarization."[53] At the core of this passage is his realization that the global interrelatedness of natural and social systems, and humanity's dependency on the ecologically sound use of resources as promoted by IUCN scientists, had not fully been recognized.

While Nicholson's skepticism was well founded, in recent years some developments suggest that there is at least some renewed theoretical interest in environmental schemes that are both global and inclusive. In the past two decades, with the accumulation of more and more scientific evidence for human-induced climate change, planetary thinking and earth systems research have received a new impetus, in many ways echoing the systems project of the 1960s and 1970s.[54] Since the year 2000, the idea that we have entered a new geological age, the Anthropocene, has received a lot of attention not only in the geophysical sciences but also in the social sciences and the humanities. The idea is based on the fact that human industrialization and modernization can be recognized in the earth's stratigraphic record. In August 2016, an official expert group recommended the acceptance of the new epoch to the International Geological Congress held in Cape Town, dating its beginning to the early years of the atomic age, around 1950.[55] The concept of the Anthropocene entails the idea that human beings have become a planetary force, and that all life on Earth, as well as all nonliving environments, are recognizably shaped by human existence. The broad acceptance of human-caused climate change has sparked new interest in earth systems science, which uses approaches from chemistry, biology, and geophysics to study the planet as a "singular, highly-interactive system."[56] Likewise, scholarship in the social sciences and humanities has increasingly been concerned with concepts—such as the Anthropocene—that try to capture a state of global interconnectedness between human societies and their environments.[57] In this way, the concept connects research in the natural sciences with ethical questions on environmental justice, postmodern discussions on human-nature relationships, and debates on nonhuman agency, responsibility, and scientific integrity, significantly broadening the disciplinary breadth of environmental expertise.[58] Also within international conservation and environmental policymaking at large, systematic thinking has taken new root; in 2015 IUCN embraced the UN's 17 Sustainable Development Goals (SDGs) and its 169 targets as "integrated and indivisible, global in nature and universally applicable,"[59] and has presented itself as key player in achieving the UN's SDGs.[60] The SDG campaigns surely suggest a move toward a global agenda.[61]

Whether global warming, with its unpredictable and dislocated consequences, might lead to a truly global and inclusive venture in the spirit of the Anthropocene remains to be seen, however. IUCN itself speaks of a "cautious optimism."[62] The ephemeral nature of international environmental commitments was reconfirmed in 2017 when the hopes for a binding climate agreement that had grown during the Obama administration were turned down by Donald Trump. While questions about global solutions and environmental expertise remain highly topical and unresolved, the history of IUCN presented in this book might help us understand where these persisting fields of tensions have come from: the intrinsic entanglement of international environmental policymaking and the development discourse of the 1970s.

Notes

1. E.g., Sabine Höhler, *Spaceship Earth in the Environmental Age, 1960–1990* (London: Pickering & Chatto, 2015); Paul Warde, Libby Robin, and Sverker Sörlin, *The Environment: A History of the Idea* (Baltimore: John Hopkins University Press, 2018).
2. In the past, it has mostly been either uncontested accomplishments or sclerotic oppositions that have brought the environmental field to the attention of scholarly investigation. For two authors who have focused on environmental success stories, see John McCormick, "The Origins of the World Conservation Strategy," *Environmental Review* 10, no. 3 (1986); Peter M. Haas, *Saving the Mediterranean: The Politics of International Environmental Cooperation* (New York: Columbia University Press, 1990). Wade Rowland describes the debate between advocates for the management of human societies such as Paul Ehrlich and the anti-population planning critique of Barry Commoner's *The Closing Circle*, which would culminate in a public dispute in 1972 at the UN Conference on the Human Environment: Wade Rowland, *The Plot to Save the World: The Life and Times of the Stockholm Conference on the Human Environment* (Madison: Clarke, Irwin, 1973); Paul R. Ehrlich, *The Population Bomb* (New York: Buccaneer Books, 1968); Barry Commoner, *The Closing Circle: Man, Nature, and Technology* (New York: Knopf, 1971). Covering forty years of disagreements about the environment, John Dryzek and David Schlosberg draw attention to the diversity of opinions represented within environmental politics: John S. Dryzek and David Schlosberg, *Debating the Earth: The Environmental Politics Reader* (Oxford: Oxford University Press, 2005). Paul Sabin recounts how, in the 1970s and 1980s, techno-optimistic economists like Julian Simon challenged Ehrlich's warnings that predicted resource scarcity and famine if no action was taken against the looming perils of human overpopulation: Paul Sabin, *The Bet: Paul Ehrlich, Julian Simon, and Our Gamble over Earth's Future* (New Haven: Yale University Press, 2014). So far, the few historical works that have looked at both the history of environmental sciences and policymaking in the 1960s and 1970s have usually paid little attention to controversies between promoters of different scientific approaches to the environmental crisis and how these related to ideas on environmental governance. Authors such as Stephen Bocking and Peter Haas, for example, have looked at science in action in the context of different environmental policy debates: Stephen Bocking, *Nature's Experts: Science, Politics, and the Environment* (New Brunswick: Rutgers University Press, 2004); Haas, *Saving the Mediterranean*. Their focus has been on how science in general was used in the decision-making process and how it is part of democracy; they have not looked at the interplay between scientific and political lines of argumentation. In practice, however, in the formulation of environmental advice and guidelines, expertise and methods are continuously moderated and adjusted.
3. For extensive accounts and discussions about the positive impact that IUCN's program has had during its seventy-year existence, see Martin W. Holdgate, *The Green Web: A Union for World Conservation* (Gland: IUCN, 1999); Raymond F. Dasmann and Randall Jarrell, *Raymond F. Dasmann: A Life in Conservation Biology* (Bloomington: Xlibris

Corporation, 2000); Maurice F. Strong, *Where on Earth Are We Going?* (Toronto: Knopf Canada, 2000); IUCN, "The Union," https://www.iucn.org/secretariat/about/union.

4. IUCN, *Results of the IUCN Programme 2005–2008* (Gland: IUCN, 2009); Carijn Beumer and Pim Martens, "IUCN and Perspectives on Biodiversity Conservation in a Changing World," *Biodiversity and Conservation* 22, no. 13–14 (2013).

5. Raymond F. Dasmann, "Stockholm, Bucharest, and Rome—What Next?," *IUCN Bulletin*, n.s., 6, no. 1 (1975): 1.

6. E.g., compare Stephen Macekura, *Of Limits and Growth: The Rise of Global Sustainable Development in the Twentieth Century* (Cambridge: Cambridge University Press, 2015), 239ff.; John McCormick, *Reclaiming Paradise: The Global Environmental Movement* (Bloomington: Indiana University Press, 1991); Donald Worster, *Nature's Economy: A History of Ecological Ideas*, Studies in Environment and History (Cambridge: Cambridge University Press, 1985), 170ff.

7. In recent years, the idea of the Anthropocene has taken root in the environmental humanities. This idea of a new stratigraphical epoch marked by human impact on the world's ecosystems has sparked literature on planetary systems, deep history, and large narratives concerned with global environmental change. E.g., see John R. McNeill and Peter Engelke, *The Great Acceleration: An Environmental History of the Anthropocene since 1945* (Cambridge, MA: Belknap Press of Harvard University Press, 2014); Dipesh Chakrabarty, "The Climate of History: Four Theses," *Critical Inquiry* 35, no. 2 (2009); Helmuth Trischler, "The Anthropocene: A Challenge for the History of Science, Technology, and the Environment," *NTM Zeitschrift für Geschichte der Wissenschaften, Technik und Medizin* 24, no. 1 (2016). Partly as a reaction to a growing number of works on global environmental regimes, a number of studies has highlighted local forms of knowledge, activism, and engagement as crucial counterparts to international efforts; e.g., see Thomas F. Thornton and Patricia M. Thornton, "The Mutable, the Mythical, and the Managerial: Raven Narratives and the Anthropocene," *Environment and Society* 6, no. 1 (2015); Marco Armiero and Lise Sedrez, eds., *A History of Environmentalism: Local Struggles, Global Histories* (London: Bloomsbury Academic, 2014).

8. Indirectly, notions of different scales of action and policy often play a role in the literature on environmental governance; e.g., see Joachim Radkau, *Nature and Power: A Global History of the Environment* (Cambridge: Cambridge University Press, 2008); Frank Biermann, "'Earth System Governance' as a Crosscutting Theme of Global Change Research," *Global Environmental Change* 17, no. 1 (2007). Cultural anthropologist Ursula Heise discusses the ideas of place and planet in more theoretical depth: Ursula K. Heise, *Sense of Place and Sense of Planet: The Environmental Imagination of the Global* (New York: Oxford University Press, 2008). Similarly, French philosopher and sociologist Bruno Latour calls for a grounded, planetary politics of Earth as an alternative to the exclusive ideologies of global interconnectivity and markets that dominated the postwar period: Bruno Latour, *Down to Earth: Politics in the New Climatic Regime* (Cambridge & Medford: Polity Press, 2018).

9. Marybeth Long Martello and Sheila Jasanoff, "Introduction: Globalization and Environmental Governance," in *Earthly Politics: Local and Global in Environmental Governance*, ed. Marybeth Long Martello and Sheila Jasanoff (Cambridge, MA: MIT Press, 2004), 6ff.

10. Miklos Udvardy, *A Classification of the Biogeographical Provinces of the World* (Morges: IUCN, 1975).

11. Raymond F. Dasmann, John P. Milton, and Peter H. Freeman, *Ecological Principles for Economic Development* (New York: John Wiley & Sons, 1973).

12. Donella H. Meadows et al., *The Limits to Growth: A Report for the Club of Rome's Project on the Predicament of Mankind* (New York: Universe Books, 1972).

13. E.g., John McHale, *The Ecological Context* (New York: G. Braziller, 1970).

14. Barbara Ward and René J. Dubos, *Only One Earth: The Care and Maintenance of a Small Planet* (New York: W. W. Norton, 1972).

15. For a history of the concept of biodiversity, see David Takacs, *The Idea of Biodiversity: Philosophies of Paradise* (Baltimore: Johns Hopkins University Press, 1996); for a description of the ecosystem services framework, see José Sarukhán and Joseph Alcamo, *Ecosystems and Human Well-Being: A Framework for Assessment; A Report of the Conceptual Framework Working Group of the Millennium Ecosystem Assessment* (Washington, DC: Island Press, 2003).

16. Evert Peeters, Joris Vandendriessche, and Kaat Wils, eds., *Scientists' Expertise as Performance: Between State and Society, 1860–1960* (London: Routledge, 2015). For the self-fashioning of experts, see Stephen Hilgartner, *Science on Stage: Expert Advice as Public Drama* (Stanford: Stanford University Press, 2000).

17. E.g., E. Max Nicholson to Gerardo Budowski, 16 January 1969, RSA SCIBP: NHM, Box 1, Folder "First GA SCIBP 1964."

18. Johan Schot and Vincent Lagendijk, "Technocratic Internationalism in the Interwar Years: Building Europe on Motorways and Electricity Networks," *Journal of Modern European History* 6, no. 2 (2008).

19. C. M. Berkeley to Tracy Philips, 16 March 1956, UNESCO IUCN: Box 502.7 Folder A 01 IUCNNR-6; Udall cited in IUCN, ed., *Proceedings of the Eighth General Assembly, Kenya 1963* (Morges: IUCN, 1964), 47.

20. The continued dominance of forms of "fortress conservation," expelling human populations from protected areas, has been discussed in much detail by Dan Brockington, *Fortress Conservation: The Preservation of the Mkomazi Game Reserve, Tanzania* (Bloomington: Indiana University Press, 2002). For a discussion of the discrepancies between the changing conservation discourse since 1945 and continued exclusive practices, see Simone Schleper and Hans Schouwenburg, "Islands and Bioregions: Global Reserve Design Models and the Making of National Parks, 1960–2000," in *Spatializing the History of Ecology: Sites, Journeys, Mappings*, ed. Raf De Bont and Jens Lachmund (New York: Routledge, 2017).

21. Holdgate, *The Green Web*, 16.

22. There is a rich historiography discussing the breaks and continuities of nature conservation networks in postcolonial territories: Jim Igoe, *The Nature of Spectacle: On Images, Money and Conservation Capitalism* (Tucson: University of Arizona Press, 2017); Elizabeth Garland, "The Elephant in the Room: Confronting the Colonial Character of Wildlife Conservation in Africa," *African Studies Review* 51, no. 3 (2008); William Adams and Martin Mulligan, *Decolonizing Nature: Strategies for Conservation in a Post-Colonial Era* (London: Earthscan, 2004); Roderick P. Neumann, "The Postwar Conservation Boom in British Colonial Africa," *Environmental History* 7, no. 1 (2002). Also see Raf De Bont, Simone Schleper, and Hans Schouwenburg, "Conservation Con-

ferences and Expert Networks in the Short 20th Century," *Environment and History* 23, no. 4 (2017).

23. David A. Munro, "Conservation and Politics," *IUCN Bulletin*, n.s., 10, no. 1 (1979): 1.
24. Ossamma M. El-Tayeb, "Internal Memorandum," to Reuben Olembo, UNEP Senior Programme Officer, 30 March 1979, cited in Holdgate, *The Green Web*, 151; Thomas F. Power and Jorge Morello, "Evaluation of Projects Contracted for Execution to the International Union for Conservation of Nature and Natural Resources as a Supporting Organization," 1979, 25, HUL PST I: Box 58, Folder 555.
25. E.g., "UNEP Governing Council. Second Session, Nairobi, March 11–22, 1974, Item 8a of the Provisional Agenda," HUL PST II: Box 65, Folder 611.
26. See e.g., Macekura, *Of Limits and Growth*, 222–23; Iris Borowy, *Defining Sustainable Development for Our Common Future: A History of the World Commission on Environment and Development (Brundtland Commission)* (London: Routledge, 2013), 174, 91.
27. Höhler, *Spaceship Earth in the Environmental Age*, 21; Fernando Elichirigoity, *Planet Management: Limits to Growth, Computer Simulation, and the Emergence of Global Spaces* (Chicago: Northwestern University Press, 1999), 4. For another critique of this notion, see Jens Ivo Engels, "Modern Environmentalism," in *The Turning Points of Environmental History*, ed. F. Uekötter (Pittsburg: University of Pittsburg Press, 2010), 129–31.
28. Holdgate, *The Green Web*, 174ff.
29. Ibid., 180.
30. Robert Prescott-Allen, *National Conservation Strategies and Biological Diversity: A Report to IUCN Conservation for Development Centre* (Gland: IUCN, 1986), 1.
31. The UK's NCS, for instance, was concerted by Max Nicholson, but with little to no international coordination. Committee of IUCN Members United Kingdom and Council for Environmental Conservation United Kingdom, *Earth's Survival: A Conservation and Development Programme for the UK* (London: Programme Organizing Committee, UK Committee for IUCN, 1981–1982, 1981).
32. Jeffrey McNeely et al., *Conserving the World's Biological Diversity* (Gland; Washington DC: IUCN, WRI, WWF-US, World Bank, 1990), 21ff.
33. Borowy, *Defining Sustainable Development for Our Common Future.*
34. David A. Munro et al., *Caring for the Earth: A Strategy for Sustainable Living* (Gland: IUCN, 1991).
35. IUCN, ed., *Proceedings of the Seventeenth Session of the General Assembly of IUCN and Seventeenth IUCN Technical Meeting: San José, Costa Rica 1–10 February 1988* (Gland: IUCN, 1988), 185.
36. For some expression of disappointment, see Mark Halle, "Building the Worldwide Union," in *The Future of IUCN: The World Conservation Union*, ed. Martin W. Holdgate and Hugh Synge (Gland: IUCN, 1993), 135.
37. Macekura, *Of Limits and Growth*, 290.
38. Lyle Glowka et al., *A Guide to the Convention on Biological Diversity* (Gland: IUCN, 1994), 4.
39. Fikret Berkes, "Rethinking Community-Based Conservation," *Conservation Biology* 18, no. 3 (2004): 624. For an elaborate discussion on community-based conservation as pursued by IUCN, see Robert Fischer et al., *Linking Conservation and Poverty Reduction: Landscapes, People and Power* (Oxon: Earthscan, 2008), 125ff.

40. Raf De Bont, "'Primitives' and Protected Areas: International Conservation and the 'Naturalization' of Indigenous People, ca. 1910–1975," *Journal of the History of Ideas* 76, no. 2 (2015); Roderick P. Neumann, "Ways of Seeing Africa: Colonial Recasting of African Society and Landscape in Serengeti National Park," *Ecumene* 2, no. 2 (1995); Bernhard Gissibl, "Die Mythen der Serengeti: Naturbilder, Naturpolitik und die Ambivalenz westlicher Um-Weltbürgerschaft in Ostafrika," *Denkanstöße. Stiftung Natur und Umwelt Rheinland-Pfalz* 10 (2013).

41. Javier Beltrán, ed., *Indigenous and Traditional Peoples and Protected Areas: Principles, Guidelines and Case Studies*, Best Practice Protected Area Guidelines Series (Cardiff: World Commission on Protected Areas [WCPA]; IUCN; Cardiff University, 1998).

42. IUCN, *Proceedings of the Members' Assembly: World Conservation Congress Honolulu, Hawai'i, United States of America, 6–10 September 2016* (Gland: IUCN, 2016), 86. For the global indigenous movement at the UN, see Andrea Muehlebach, "'Making Place' at the United Nations: Indigenous Cultural Politics at the UN Working Group on Indigenous Populations," *Cultural Anthropology* 16, no. 3 (2001).

43. E.g., see Regina Scheyvens, Glenn Banks, and Emma Hughes, "The Private Sector and the SDGs: The Need to Move Beyond 'Business as Usual,'" *Sustainable Development* 2016, no. 24 (2016); Emmanuel Kumi, Albert A. Arhin, and Thomas Yeboah, "Can Post-2015 Sustainable Development Goals Survive Neoliberalism? A Critical Examination of the Sustainable Development–Neoliberalism Nexus in Development Countries," *Environment, Development and Sustainability* 16, no. 3 (2014); Bram Büscher, Wolfram Heinz Dressler, and Robert Fletcher, *Nature Inc.: Environmental Conservation in the Neoliberal Age* (Tucson: University of Arizona Press, 2014), 11.

44. Roderick P. Neumann, *Imposing Wilderness: Struggles over Livelihood and Nature Preservation in Africa* (Berkeley: University of California Press, 1998); Bernhard Gissibl, Sabine Höhler, and Patrick Kupper, *Civilizing Nature: National Parks in Global Historical Perspective* (New York: Berghahn Books, 2012).

45. Brockington, *Fortress Conservation*, 10.

46. Roger A. Pielke, *The Honest Broker: Making Sense of Science in Policy and Politics* (Cambridge: Cambridge University Press, 2007), 78ff.; Sheila Jasanoff, "Speaking Honestly to Power. Review of *The Honest Broker: Making Sense of Science in Policy and Politics*, by Roger A. Pielke," *American Scientist* 96, no. 3 (2008): 242. For a discussion of how current climate science often forgets that climate change is "at once a moral and an epistemic undertaking," see Sheila Jasanoff, "A New Climate for Society," *Theory, Culture & Society* 27, no. 2–3 (2010): 238ff.

47. Silke Beck, "Moving Beyond the Linear Model of Expertise? IPCC and the Test of Adaption," *Regional Environmental Change* 11, no. 2 (2011): 308; Kari De Pryck and Krystel Wanneau, "(Anti)-Boundary Work in Global Environmental Change Research and Assessment," *Environmental Science and Policy* 77, no. 1 (2017).

48. Oliver Milman, "Donald Trump Picks Climate Change Skeptic Scott Pruitt to Lead EPA," *The Guardian*, 8 December 2016, accessed 24 March 2019, https://www.theguardian.com/us-news/2016/dec/07/trump-scott-pruitt-environmental-protection-agency. For a critical discussion on how positivist forms of knowledge remain dominant within the fields of geography, policy planning, and natural resource management, see Edward A. Morgan and Natalie Osborne, "It's the Lungfish, Stupid: Knowledge Fights, Activism, and the Science-Policy Interface," *Geographical Research* 54, no. 4 (2016). For a

nuanced discussion about the changing perception of expertise within the field of STS, see Reiner Grundmann, "The Problem of Expertise in Knowledge Societies," *Minerva* 55, no. 1 (2017).

49. Both membership and authorship within the IPCC remains dominated by natural scientists from the United Kingdom and the United States. Esteve Corbera et al., "Patterns of Authorship in the IPCC Working Group III Report," *Nature Climate Change* 6, no. 1 (2016); David G. Victor, "Embed the Social Sciences in Climate Policy," *Nature* 520, no. 7545 (2015). The work of the Swiss researcher and artist Susan Schuppli has thematized this conflict between IPCC scientists and the Arctic indigenous community in her project "Can the Sun Lie," which was part of a larger European Research Council project at Goldsmith University called *Forensic Architectures*, concerned with the use of architectural and media evidence in cases of contemporary conflicts, including those on environmental justice. Anselm Franke, Eyal Weizman, and Haus der Kulturen der Welt, "Forensis: The Architecture of Public Truth," (Berlin: Sternberg Press, 2014), 56–65.

50. For struggles for scientific credibility and scientific boundary work, see Thomas F. Gieryn, *Cultural Boundaries of Science: Credibility on the Line* (Chicago: University of Chicago Press, 1999), e.g., 15.

51. In Washington, but also in the Netherlands, the United Kingdom, France, Germany, Ireland, Switzerland, and Belgium, scientists held Marches for Science on Earth Day (22 April) 2017. E.g., see Andrew Griffin, "Scientists to Oppose Donald Trump in Huge 'March for Science' in Washington," *The Independent*, 27 January 2017, accessed 3 March 2018, http://www.independent.co.uk/news/science/donald-trump-science-march-washington-climate-change-global-warming-a7547206.html; Lindzi Wessel, "Hundreds Rally for Science at Demonstration near AAAS Meeting," *Science* (2017), accessed 3 March 2018, http://www.sciencemag.org/news/2017/02/hundreds-rally-science-demonstration-near-aaas-meeting. Others refused to participate, seeing the marches as a politically biased activity that might counteract their image as neutral researchers; e.g., see Michael Roston, "The March for Science: Why Some Are Going, and Some Will Sit Out," *The New York Times*, 17 April 2017, accessed 3 March 2018, https://www.nytimes.com/2017/04/17/science/march-for-science-voices.html.

52. While the mission of the U.S. organization behind the marches claims to support an integrative vision for the future, there is little explanation of what this is supposed to entail, though they suggest that more evidence-based research directly leads to better policymaking. Their emphasis on the North American STEM subjects (science, technology, engineering, and mathematics) suggests that research in the social sciences and humanities is not part of their core concern. March for Science, "AAAS Advocacy Toolkit," n.d., accessed 23 March 2019, https://aas.org/advocacy-resources/reference-guide-how-advocate-science. Participating in the March for Science in Maastricht on 22 April 2017, I was slightly disappointed by its lack of interdisciplinarity and its exclusive rhetoric, seemingly based on an assumed need to defend the hard sciences from all forms of social critique.

53. E. Max Nicholson, "Unpublished Autobiography," n.d., ca. 2000–2003, AL EMN: Boxes A.7–A.12.

54. Maslin and Lewis, "Anthropocene."

55. Damian Carrington, "The Anthropocene Epoch: Scientists Declare Dawn of Human-Influenced Age," *Guardian*, 29 August 2016, accessed 23 March 2019, https://

www.theguardian.com/environment/2016/aug/29/declare-anthropocene-epoch-experts-urge-geological-congress-human-impact-earth. Of the thirty-five members of the international Anthropocene Working Group (AWG), which has been active since 2009, thirty-four voted for the end of the Holocene, the last official Global Standard Stratigraphic Age, which began about twelve thousand years ago when the first human settlements formed. One AWG member abstained from the vote. AWG, "Working Group on the 'Anthropocene,'" Subcommission on Quaternary Stratigraphy, n.d., accessed 23 March 2019, http://quaternary.stratigraphy.org/workinggroups/anthropocene/; Jan Zalasiewicz and Colin Waters, "Media Note: Anthropocene Working Group (AWG)," University of Leicester Press Releases, 29 August 2016, accessed 23 March 2019, http://www2.le.ac.uk/offices/press/press-releases/2016/august/media-note-anthropocene-working-group-awg.

56. Stanford University's School of Earth, Energy and Environmental Sciences, "Research," n.d., accessed 7 November 2016, https://pangea.stanford.edu/ess/research.

57. E.g., Mark A. Maslin and Simon L. Lewis, "Anthropocene: Earth System, Geological, Philosophical and Political Paradigm Shifts," *The Anthropocene Review* 2, no. 2 (2015).

58. For excellent, interdisciplinary discussions of the Anthropocene epoch, see Clive Hamilton, "Human Destiny in the Anthropocene," in *The Anthropocene and the Global Environmental Crisis: Rethinking Modernity in a New Epoch*, ed. Clive Hamilton, François Gemenne, and Christophe Bonneuil (Abington-on-Thames: Earthscan, 2015), 32–42; Chakrabarty, "The Climate of History"; Manfred D. Laubichler and Jürgen Renn, "Extended Evolution: A Conceptual Framework for Integrating Regulatory Networks and Niche Construction," *Journal of Experimental Zoology Part B: Molecular and Developmental Evolution* 324, no. 7 (2015); Jürgen Renn and Manfred D. Laubichler, "Extended Evolution and the History of Knowledge," in *Integrated History and Philosophy of Science: Problems, Perspectives, and Case Studies*, Vienna Circle Institute Yearbook, ed. Friedrich Stadler (Vienna: Springer Int'l., 2017); Jürgen Renn, Manfred D. Laubichler, and Helge Wendt, "Energy Transformations between Coffee and Coevolution," in *Welcome to the Anthropocene: The Earth in Our Hands. A Special Exhibition by the Deutsches Museum and the Rachel Carson Center for Environment and Society*, ed. Nina Möllers, Christian Schwägerl, and Helmuth Trischler (Munich: Deutsches Museum, 2014).

59. IUCN, *Draft International Covenant on Environment and Development—Implementing Sustainability—Fifth Edition: Updated Text* (Gland: IUCN, 2015), 47.

60. IUCN, *IUCN Programme 2017–2020. Approved by the IUCN World Conservation Congress, September 2016* (Gland: IUCN, 2016).

61. In 2014, Ban Ki-moon, then secretary-general of the UN, called the SDGs a "paradigm shift for people and the planet." Ban Ki-moon, *The Road to Dignity by 2030: Ending Poverty, Transforming All Lives and Protecting the Planet; Synthesis Report of the Secretary-General on the Post-2015 Agenda* (New York: UN, 2014), 7. Also see Jeffrey D. Sachs, "From Millennium Development Goals to Sustainable Development Goals," *The Lancet* 379, no. 9832 (2012).

62. IUCN, *IUCN Programme 2017–2020*, 2.

 APPENDIX

Expert Biographies

Harold Jefferson Coolidge (1904–1985)

Harold Jefferson Coolidge was born in Boston, Massachusetts, in 1904, as a third great-grandson of Thomas Jefferson, third president of the United States. Already in his early twenties, Coolidge developed a strong interest in nationally and internationally organized nature conservation, first in New England, later also abroad. In the mid-1920s, Coolidge joined the American Boone and Crockett Club, a prestigious society for hunting and habitat preservation founded by Theodore Roosevelt in 1887. Interested in both natural history and hunting, Coolidge was fascinated by the stories delivered at the club by renowned guests and speakers, such as the American naturalist and Africa traveler Frederick Russell Burnham.

For a more formal training in the natural sciences, Coolidge first went to Milton Academy and the University of Arizona, then to study for a science undergraduate degree at Harvard College, which he received in 1927. As a Harvard College student, Coolidge entered into an affiliation with the Harvard Museum of Comparative Zoology and undertook his first trip to the African continent himself, joining the Harvard Medical Expedition to Africa in 1926 and 1927. Traveling through Liberia and the (then) Belgian Congo, Coolidge specialized in primatology, which would remain his concentration during the course of his studies. Upon his return, Coolidge commenced a zoology graduate degree in Cambridge, England. He was soon asked by the Field Museum of Chicago to join a second expedition to the Mekong River in (then) Indochina, planned for the years 1928 and 1929. During this trip, Coolidge developed an interest in the natural environment of the Pacific region at large, which would stay with him for the rest of his life. The participation in expeditions, often to the European colonies that were less developed and believed to still contain more genuine natural environments, was typical for conservation-minded biologists of his generation. Many of Coolidge's colleagues at IUCN developed their conservation interests while studying ecosystems far removed from their home universities.[1]

In 1929, Coolidge had gathered enough credentials to join the Museum of Comparative Zoology at Harvard as assistant curator of mammals and associate in mammalogy. The years he spent at the museum were important for his conservation career as well as his personal life. In the summer of 1929, Coolidge joined the Bureau of Biological Survey in Alaska for a few months, a trip during which he gathered firsthand experience in the management of large scientific projects. First conservation efforts followed. In 1930, Coolidge established the American Committee for International Wildlife Protection, together with several of his peers from the Boone and Crockett Club. A year later, in 1931, Coolidge married Helen Carpenter Isaacs, with whom he had three children: Nicholas Jefferson, born in 1932, Thomas Richards, born in 1934, and Isabelle Gardner, born in 1939.

Coolidge's time at the Museum for Comparative Zoology came to an end in 1942. With the United States' involvement in World War II, Coolidge joined the Office of Special Services, first in Washington, then in England, France, and Italy. After this second stay in Europe, Coolidge returned to the United States once more, resuming his scientific career, this time at the U.S. National Academy of Sciences. Soon he helped organize the First Pacific Science Congress, taking on the position as executive director of the National Academy's Pacific Science Board in 1946. In this function, Coolidge attended the 1948 Fontainebleau Conference, which led to the foundation of the International Union for the Preservation of Nature (IUPN). He was elected vice president of IUPN at the same meeting. As vice president, Coolidge effectuated several decisions that shaped and formed IUPN and the renamed International Union for Conservation of Nature and Natural Resources (IUCN) during the coming years. His initiatives included the foundation of the Species Survival Commission (SSC), which he chaired from 1950, and the Commission on National Parks, which he led from 1958, first as a working group and then as a full commission from 1960 onward. Coolidge played an important role in the Union's early international achievements, such as the First World Congress on National Parks, convened in Seattle in 1962, and early publications of the IUCN's *List of National Parks and Protected Areas*.

Coolidge helped to establish some of IUCN's most significant interinstitutional links. In 1961, for instance, he was one of the founding fathers of the World Wildlife Fund (WWF), and he remained an important intermediary between the two organizations for the rest of his life. Due to these achievements, Coolidge was elected president of IUCN in 1966, a post he held for six years, until after the Stockholm Conference of 1972. From 1972, he remained active as honorary president of IUCN. In the same year, he divorced Helen Carpenter Isaacs and married Martha (Muffy) Thayer Henderson, who actively supported Coolidge in his honorary offices at IUCN and the WWF International Board, which he held until 1978.

Coolidge remained active in international conservation efforts into old age. In 1980, he received the Paul Getty Wildlife Conservation Prize for his lifetime effort. In 1983, he sponsored the Coolidge Center for Environmental Leadership in Cambridge, Massachusetts, a specialist training center for environmental management that flourished into the 1990s. Coolidge died in 1985 in Beverly, Massachusetts. He rests at Monticello, Thomas Jefferson's home in Virginia.[2]

Edward Max Nicholson (1904–2003)

Like Coolidge, Max Nicholson was born in 1904. As an Englishman, he represented another strong nation in international nature conservation efforts of the twentieth century. As a boy, Nicholson went to Sedbergh School in Cumbria and, from 1926 onwards, studied history at Hertford College, Oxford, being one of the few leading IUCN conservationist without a degree in the life sciences. Growing up in the Irish and English countryside, he became an enthusiastic ornithologist at a young age, publishing his first books on birds in 1926 and 1927, and co-organizing two bird censuses during his time at Oxford. His interest in the natural world was not confined to ornithology alone, however. Looking for likeminded peers, Nicholson soon made friends with biology students and attended lectures by renowned ecologists, such as Julian Huxley and Charles Elton. Eager to share his fascination for the natural world, in 1927, Nicholson founded the Oxford Exploration Club to support and encourage original expeditions planned and undertaken by Oxford students. He himself took part in several research trips—for instance, to Greenland in 1928 and to the Amazonian rain forest in British Guiana in 1929.

After graduating, Nicholson first worked in publishing, joining the London magazine *Week-End Review* as an assistant editor in 1931. While not yet pursuing a career as a naturalist, his newspaper job allowed him to spread his ideas and to publish his first environmentalist essays. In 1931, Nicholson drafted *A National Plan for Britain*, a manifesto for the organization Political and Economic Planning (PEP), which he, together with Huxley and a handful of likeminded societal planners and naturalists, had established the same year. Among other things, the centralized planning institute was influential in the formation of the National Health Service and supported several British overseas development projects.[3] Nicholson never lost his interest in birds and in 1932 created the British Trust for Ornithology. Yet, more profoundly than their American peers, young British naturalists like Nicholson were soon affected by the onset of war. After the outbreak of World War II, PEP focused on natural resource management within Britain, and Nicholson attained a related position at the Ministry of War Transport.[4]

After the war, Nicholson first held a government position in the reconstruction efforts, but retained his interest in nature conservation and natural resource planning and soon joined the British Nature Conservancy, which by then shared its headquarters with PEP. In 1952, he was elected the Conservancy's director general, a position he held until 1966. Like Coolidge, Nicholson had been influential at the Fontainebleau Conference in 1948 and played a significant role in the establishment of the WWF in 1961. From 1964 onward, Nicholson headed the Section on the Conservation of Terrestrial Communities (IBP/CT) of the International Biological Program (IBP). In 1966, Nicholson left the Nature Conservancy to found his own landscape planning bureau, Land Use Consultants, in London. After the end of the IBP in 1974, Nicholson remained active in the Royal Society for the Protection of Birds, of which he served as president in the 1980s. During the same period, he was elected vice-president of the U.K. branch of the WWF, a position he held until his death. Like Coolidge, Nicholson won several awards, among them the WWF Gold Medal in 1982. In 1995, Nicholson set up his last big project, a think tank called the New Renaissance Group, which summoned together a small group of his former colleagues to revive the discussions on conservation and development that he had championed in the 1960s and 1970s.

Nicholson was married twice. His first wife, Mary Crawford, died in 1995, and his second wife, Marie Mauerhofer, known as Toni, died in 2002. From the two marriages he had three sons, Piers, Tom, and David. Nicholson passed away from old age in 2003, outliving many of his contemporaries at IUCN by more than a decade.[5]

Raymond Dasmann (1919–2002)

Born in 1919 in San Francisco, Raymond Dasmann belonged to a slightly younger cohort of IUCN members compared to Coolidge and Nicholson. Dasmann attended school and college in San Francisco, but, before finishing his education, he joined the U.S. Army in 1942. As part of his military service, he traveled to New Guinea and Australia, where he met his wife, Elizabeth Sheldon, with whom he would have three daughters. After the war, the couple returned to San Francisco, and Dasmann went to University of California, Berkeley. Here he spent the following eight years completing an undergraduate degree in 1948, a master's degree in 1951, and a doctorate in zoology in 1953. As such, Dasmann was part of a generation of conservation-minded biologists who completed their degrees after their wartime involvement, in the postcolonial period. For his Ph.D., Dasmann focused on Northern California deer populations under the supervision of A. Starker Leopold. From there, Dasmann moved to further research and teaching jobs, first at the University of Minnesota from 1953 to 1954, and then to Humboldt State University in Arcata, California, where he stayed until 1965, teaching range ecology and management.[6]

For Coolidge and Nicholson, expeditions to the European colonies on the African and Asian continents had been part of their university education. Dasmann also had the chance to gather some field experience abroad as part of his early work in postwar California. Ecological and zoological research stays abroad were by now, however, less adventure seeking, and had taken on a more structured, academic form. In 1959, during his stay at Humboldt State, Dasmann went to Southern Rhodesia (Zimbabwe) on a Fulbright Fellowship to work on threatened African wildlife and to collect positive conservation examples by indigenous groups, an experience that would continue to influence the rest of his conservation career.

In 1966, Dasmann left academia for an appointment as director of international programs at the Conservation Foundation in Washington, DC. He joined IUCN in the position of senior ecologist the same year. In both functions, Dasmann was involved in advising the United Nations Educational, Scientific and Cultural Organization (UNESCO) during their preparations for the Man and the Biosphere Program (MAB), and in preparing for IUCN's participation in the Stockholm Conference of 1972. In 1977, after internal discrepancies regarding the management of IUCN, Dasmann left his post there and returned to his home state to accept a position at the University of California, Santa Cruz. In his new position as professor of environmental studies, which he held until his retirement in 1989, Dasmann was an influential figure in the establishment of the new research field of conservation biology, influenced by new research in biogeography and a more activist stance to conservation policies.[7] Throughout this period, he nevertheless continued publishing for IUCN, served as a consultant for various local, national, and international ecology-based conservation and development projects, and was on the board of directors of Friends of the Earth. After the death of his wife in 1996, Dasmann dedicated his time to the creation of the Golden Gate Biosphere Reserve in Northern California. He passed away in 2002 in Santa Cruz after several years of illness.

Gerardo Budowski (1925–2014)

Born in 1925 Berlin to a Russian father of Jewish background and a German mother, Gerardo Budowski had relocated with his parents to France in 1938, and from there to Venezuela after the German occupation of France in 1940. His displaced childhood formed the backdrop to an outstandingly international career in ecology and conservation. Budowski earned his first university degree in Venezuela as an agricultural engineer from the Central University in Caracas in 1948. In 1954, he received a master's degree from the Inter-American Institute for Cooperation on Agriculture in Turrialba, Costa Rica, and a Ph.D. from the Yale School of Forestry in 1962. Afterward, he returned to

Turrialba to head the Natural Resource Program. Here he met his wife, Thelma Palma, a fellow teacher, with whom he would have two daughters. In 1967, Budowski spent a year as visiting professor at the University of California, Berkeley. Between 1964 and 1974, the forestry and mountain ecology expert, with his strong ties to the conservation community in Latin America and tremendous language skills, was an important figure within the IBP/CT section. In parallel, he was appointed head of Ecology and Conservation at UNESCO in 1967, where he contributed majorly to the preparations of the Paris Biosphere Conference of 1968. In 1970, Budowski moved from UNESCO to IUCN to serve as its first director general, a position he held until 1976.[8] Together with Coolidge, Budowski lead IUCN's efforts at the Stockholm Conference in 1972.

In 1976, Budowski's term came to an end after an internal struggle within IUCN's administrative ranks after the General Assembly in Kinshasa, Zaire (present-day Congo), which was organized on the premises of Mombuto Sese Seko's residence, and supposed to strengthen the ties between European and American conservationists and government officials of countries in the global South. Conceptual differences, poor organization, and an uncritical attitude toward the dictator dismayed large numbers of conferees, and in the end Budowski was asked not to run for a second term.[9]

After his stepping down, Budowski continued to serve as special advisor to the WWF and the World Bank. In 1976, Budowski returned to the renamed Tropical Agricultural Research and Higher Education Center in Turrialba, Costa Rica, where he directed the Natural Resource Department until 1986. He then joined the United Nations (UN) University for Peace in Costa Rica, as director of renewable resources, and from 1989 to 1996, held a position with the technical advisory committee of the Consultative Group on International Research (CGIAR). Throughout this period, he stayed an honorary member of both IUCN and the WWF.

In 1990, Budowski received the Fred Packard Award by the World Commission on Protected Areas (formerly Commission for National Parks and Protected Areas) for his outstanding service to protected areas worldwide. The founding father of forest ecology (and enthusiastic chess player) stayed active in conservation into his eighties. In 2011, he was paid tribute by the IUCN Regional Office for Mexico, Central America and the Caribbean for his career and life dedicated to conservation. Three years later, in 2014, Budowski passed away.[10]

Martin Edward Duncan Poore (1925–2016)

Duncan Poore came to IUCN and international nature conservation through a scientific career in botany. First, however, he, like Nicholson, pursued a degree in the humanities. Dasmann commenced his studies at Trinity College in Glenalmond, Scotland. He then went to Cambridge to study classics. His studies were interrupted, however, by World War II, and, after returning to Cambridge in 1947, he switched to the natural science tripos. Poore continued his studies to pursue a doctoral degree in botany, using aerial photography to study the vegetation of Cambridgeshire. After obtaining his Ph.D., Poore traveled to the Middle East, Cyprus, and Pakistan, conducting several botanical studies, employing and refining the same methodology for aerial observation. Having gathering significant field experience, Poore was offered an attractive position at the University of Malaya, as professor of botany and dean of science. He and his wife, Judy, with whom he had two sons, Robin and Alasdair, relocated to Malaysia. From there, the couple went on several expeditions that would influence Poore's later writings on the biological diversity of natural ecosystems.

In 1965, the Poores returned to Britain after Duncan had been offered a position at Oxford University. After one year at Oxford, Poore took over from Nicholson as director of the British Nature Conservancy. As such, Poore was part of the IBP/CT and of the scientific follow-up program, the UNESCO-led MAB. At about the same time, in 1966, he joined IUCN's executive board. In 1973, Poore took over the position of IUCN senior ecologist, previously held by Dasmann, and the task to work on scientific guidelines for the conservation of tropical forests, mountains, and arid lands. Based on his scientific achievements, he was made the Union's scientific director a few years later and was assigned the function of acting director general in the years 1976 and 1977. In 1978, Poore joined UNESCO but continued to publish and advise on behalf of IUCN. He was one of the main drafters of the sourcebooks accompanying the different drafts of the *World Conservation Strategy*, published in 1980 as a joined venture by IUCN, UNESCO, the United Nations Environmental Program (UNEP), and the United Nation's Food and Agricultural Organization (FAO).[11]

In 1980, Poore returned to Oxford for a position as professor of forest science and director of the Commonwealth Forestry Institute. After a few years, however, struggling to receive sufficient financial assistance for the newly founded institute, he resigned and became an independent advisor and consultant with UNESCO, the World Bank, and the International Institute for Environment and Development (IIED). At the IIED, he set up the Forestry and Land Use Programme. During the 1980s, he also worked as a consultant for the International Tropical Timber Organization (ITTO), advising on sustainable development in the forestry industry. In 1987 at ITTO, Poore performed the first survey of sustainable forest management in timber producing countries. Numerous missions on sustainable forest management—to Sarawak in Malaysia and several places in Bolivia—followed during the late 1980s and the 1990s. In the early 2000s, Poore became a member of Nicholson's New Renaissance

Group. He continued publishing on botany and sustainable development into the 2010s. Poore passed away in April 2016 in Inverness, Scotland.[12]

Maurice Frederick Strong (1929–2015)

In his life, Strong, born in 1929 in Oak Lake, Manitoba, Canada, became one of the most renowned advocates for international environmental policy. Coming from a background quite different from that of most IUCN conservationists, Strong wandered a career path that led him to the environment and to conservation through earlier stints in business and diplomacy. Strong had left home at the age of fourteen to become a merchant and a fur trader in the Canadian North.[13] After the war, however, he befriended Noel Monod, treasurer of the UN, who sparked Strong's interest in the work of international organizations. With Monod's help, Strong was in line for a junior clerical role at the UN's Palestine Commission in late 1948. Yet, a few months prior to Strong's planned appointment, Folke Bernadotte, the Swedish mediator in the Arab-Israeli conflict, was assassinated, and the mission was consequentially restructured. From 1951, Strong resumed building up a business career for himself at the Canadian oil company Dome Petroleum. For Dome Petroleum, Strong soon became an oil scout and, together with his first wife, Pauline Olivette, traveled the Middle East and several African countries. Returning to Canada in 1955, Strong first worked for Dome, then for Petro-Canada, and from 1968 for the Canadian Development Investment Corporation and the Canadian International Development Agency (CIDA).[14]

His job at CIDA allowed Strong to come in on diplomatic, development, and resource issues in the running up to the UN Stockholm Conference of 1972, when the Swedish government sought his advice on how to stimulate the participation by governments from the global South and the Eastern Bloc in the conference and its preparations. His diplomatic abilities led to his appointment as secretary general of the conference and first executive director of UNEP, a position he retained until 1975. In the same period, he was elected to the WWF's board of trustees. Strong had already made first contact with IUCN in 1971 during the preparatory phase of the Stockholm Conference. In 1977, after the troublesome period in Kinshasa and an administrative restructuring process within IUCN, Strong moved into the organization as chairman of the newly formed Bureau to the Council, the equivalent to a board of directors. Strong held the position of IUCN patron for more than fifteen years, remaining influential in the background of several decisive moments in the Union's history.

In the 1980s, Strong, now married to his second wife, Hanne Marstrand, became a member of the World Commission on Environment and Development, also known as the Brundtland Commission, deliberating the UN's approach to sustainable development. Once more, he served as secretary general of a UN environmental conference when, in 1992, the Conference on Environment and Development, better known as the Rio Earth Summit, was held in Rio de Janeiro. Alongside his affiliation with IUCN, Strong stayed an influential figure in UN development politics throughout the 1990s, serving as advisor to, among others, UN Secretary-General Kofi Annan, World Bank president James Wolfensohn, and the World Economic Forum. Like Budowski, in 1999 Strong worked for a short period at the UN University for Peace. He remained involved in UN environmental conferences, publishing on issues related to international politics, sustainable development, and climate change into the 2010s. Strong passed away in November 2015 in Ottawa, Canada. From his two marriages, Strong had seven children.[15]

Notes

1. Walter Sullivan, "Harold Coolidge, Expert on Exotic Mammals," *The New York Times*, 16 February 1985, 337; Anna-Katharina Wöbse, "'The World after All Was One': The International Environmental Network of UNESCO and IUPN, 1945–1950," *Contemporary European History* 20, no. 3 (2011).
2. David Hughes-Evans, "Interview: Profile of Harold Jefferson Coolidge," *The Environmentalist* 1, no. 1 (1981).
3. See Alison Bashford, *Global Population: History, Geopolitics, and Life on Earth* (New York: Columbia University Press, 2014), 168ff.; Eugene Grebenik, "World Population and Resources," *Political Quarterly* 26, no. 4 (1955); Matthew Hilton, *Consumerism in Twentieth-Century Britain: The Search for a Historical Movement* (Cambridge: Cambridge University Press, 2003), 99ff.; Harold L. Smith, *War and Social Change: British Society in the Second World War* (Manchester: Manchester University Press, 1990), 64ff.
4. Robert Boote, "Obituary. Max Nicholson: The Prime Mover of the Nature Conservancy and the World Wildlife Fund, Who Helped Inspire Nature Reserves and Ecological Research," *The Guardian*, 28 April 2003, accessed 24 March 2019, http://www.theguardian.com/news/2003/apr/28/guardianobituaries.highereducation.
5. An unpublished autobiography, written by Nicholson in the early 2000s, is available at the Alexander Library of Ornithology at Oxford University. E. Max Nicholson, "Unpublished Autobiography," n.d., ca. 2000–2003, AL EMN: Boxes A.1–A.12.
6. Elaine Woo, "Raymond F. Dasmann, 83: A Founding Father of Environmentalism," *Los Angeles Times*, 9 November 2002, accessed 18 October 2018, http://articles.latimes.com/2002/nov/09/local/me-dasmann9.
7. The story of Santa Cruz–based conservation biology and Dasmann's role in the establishment of the field has been captured in an extensive oral history project, conducted in 2000 by Randall Jarrell. Raymond F. Dasmann and Randall Jarrell, *Raymond F. Dasmann: A Life in Conservation Biology* (Bloomington: Xlibris Corporation, 2000).
8. Fausto Sarmiento, "Gerardo Budowski: A Beacon to Conservation of Tropical Mountains," *Mountain Research and Development* 22, no. 2 (2002).

9. Former IUCN President Martin Holdgate has described this incident in his chapter on "The Night of the Long Knives," in Martin W. Holdgate, *The Green Web: A Union for World Conservation* (Gland: IUCN, 1999), 124ff.

10. Society of American Foresters, "Obituary: Budowski, Gerardo (1925–2014)," *International Forestry Working Group Newsletter*, December 2014, accessed 24 March 2019, http://www.orrforest.net/saf/Dec2014.pdf.

11. Roger Crofts and Des Thompson, "Professor Duncan Poore: Global Conservationist," *Scottish Natural Heritage*, 12 April 2016, accessed 24 March 2019, https://scotlandsna ture.wordpress.com/2016/04/12/professor-duncan-poore-global-conservationist/.

12. IIED, "Remembering Duncan Poore, a Leading Light in Sustainable Forestry," published electronically, 19 October 2016, accessed 24 March 2019, https://www.iied.org/ remembering-duncan-poore-leading-light-sustainable-forestry.

13. Wade Rowland, *The Plot to Save the World: The Life and Times of the Stockholm Conference on the Human Environment* (Madison: Clarke, Irwin, 1973), 35–37.

14. Maurice F. Strong, *Where on Earth Are We Going?* (Toronto: Knopf Canada, 2000), 87ff.

15. Ronald Bailey, "Who Is Maurice Strong?," *National Review*, 1 September 1997; Sam Roberts, "Maurice Strong, Environmental Champion, Dies at 86," *New York Times*, 1 December 2015, accessed 24 March 2019, http://www.nytimes.com/2015/12/02/world/ americas/maurice-strong-environmental-champion-dies-at-86.html?_r=0.

Bibliography

Archival Collections

Aberdeen

University of Aberdeen Special Collections:
International Biological Programme Papers and Publications, collected by Dr. E. M. Nicholson, MS 3162, Boxes 1–31, ca. 1961–1980 (UASC IBP).

Amsterdam

Bibliotheek van het Koninklijk Instituut voor de Tropen:
UNEP Annual Review 1975–1980, UNEP Governing Council reports 1975–1980 (KIT).

Cambridge, MA

Environmental Science and Public Policy Archives, Harvard University Library:
Maurice F. Strong Papers, Boxes 27, 28, 33, 35, 41, 42, ca. 1970–1982 (HUL MFS).

Environmental Collection of Peter S. Thacher, part I, Boxes 25, 46, 58, 64, 65, 67, 71, 95, 102, ca. 1970–1985 (HUL PST I).

Environmental Collection of Peter S. Thacher, part II, Boxes 11, 14, 15, 18–21, 25, 42, 44, 52, ca. 1970–1985 (HUL PST II).

Harvard University Archives:
Papers of Harold Jefferson Coolidge, HUG (FP), sections 78.10, 78.14, 78.16, 78.19, 78.20, 78.75, ca. 1945–1985 (HUA HJC).

Cambridge, UK

Cambridge University Special Collections:
Papers and Correspondence of Sir Peter Markham Scott F. R. S., NCUACS 87/8/99, Boxes C.651–C. 1066, ca. 1948–1990 (CUSC PMS).

Gland

IUCN Library:
IUCN Bulletin (new series) 1970–1982, Environmental Conservation 1974–1985, IUCN Yearbooks 1972–1976, sourcebooks to the World Conservation Strategy 1978–1980, National Conservation Strategies 1980–1985 (IUCNL).

London

Linnean Society Archives:
Papers of Edward Max Nicholson Related to the International Biological Program, 18 Boxes, ca. 1963–1982 (LSA EMN/IBP).

Royal Geographical Society Archives:
Papers and Correspondence of Edward Max Nicholson, Boxes 1–4, ca. 1928–1968 (RGSA EMN).

Royal Society Archives:
Papers of the Special Committee for the International Biological Program, NHM, Boxes 1, 5, 12, 13, 71, 72, 75, 79, 94, ca. 1961–1975 (RSA SCIBP).

Oxford

Alexander Library for Ornithology, Oxford University:
E. M. Nicholson Papers, Boxes A.8–A. 12, C.496–C.504, ca. 1970–2003 (AL EMN).

Weston Library, Oxford University:
Papers of Richard Fitter, Boxes 39–41, 56, ca. 1973–1983 (WL RF).

Paris

UNESCO Archives:
International Union for Conservation of Nature and Natural Resources, Box 502.7, Folder A 01 IUCNNR-6, ca. 1950–1975 (UNESCO IUCN).

Interviews and Written Communication

Douglas, Gina, former Scientific Director of the IBP/CT Section, interviewed on 19 October 2013 in London. Interviewer: Simone Schleper.

Norman, Jennifer, former secretary of Max Nicholson for the IBP/CT Section, interviewed on 8 March 2014 in London. Interviewer: Simone Schleper.

Norman, Jennifer, "Conservation with Dr. Poore," email sent to Simone Schleper on 13 March 2014.

Talbot, Lee M., former senior ecologist and chairman of IUCN's Species Survival Commission, interviewed on 11 September 2014 via Skype. Interviewer: Raf de Bont.

Talbot, Lee M. "Interview Questions," email sent to Simone Schleper on 7 July 2016.

Published Sources

Aaronson, Terri. "World Priorities." *Environment: Science and Policy for Sustainable Development* 14, no. 6 (1972): 4–13.

Adam, Rachelle. *Elephant Treaties: The Colonial Legacy of the Biodiversity Crisis.* Lebanon, NH: University Press of New England, 2014.

————. *Nature, Colonialism and Conservation Organizations: How International Law Became the Response to the Biodiversity Crisis.* Jerusalem: Hebrew University of Jerusalem, 2012.

Adams, Alexander B., IUCN, and UNESCO, eds. *First World Conference on National Parks: Proceedings of a Conference, Seattle, Washington, June 30–July 7, 1962.* Washington, DC: National Park Service, U.S. Department of the Interior, 1964.

Adams, William M. *Against Extinction: The Story of Conservation.* London: Earthscan; Fauna & Flora International, 2004.

————. *Green Development: Environment and Sustainability in a Developing World.* London: Routledge, 2008.

————. *Green Development: Environment and Sustainability in the Third World.* London: Routledge, 2001.

Alcamo, Joseph, and Elena M. Bennett. *Ecosystems and Human Well-Being: A Framework for Assessment.* Washington, DC: Island Press, 2003.

Allen, Robert. "Viewpoint. Interdependence: The Trend We Cannot Buck." *IUCN Bulletin,* n.s., 6, no. 3 (1975): 9–10.

Andersen, Mikael S., and J. Duncan Liefferink. *European Environmental Policy: The Pioneers.* Manchester: Manchester University Press, 1997.

Andersen, Stephen O., K. Madhava Sarma, and Lani Sinclair. *Protecting the Ozone Layer: The United Nations History.* London: Earthscan, 2012.

Anker, Peder. "Buckminster Fuller as Captain of Spaceship Earth." Minerva 45, no. 4 (2007): 417–34.

————. "The Ecological Colonization of Space." *Environmental History* (2005): 239–68.

————. *Imperial Ecology: Environmental Order in the British Empire, 1895–1945.* Cambridge, MA: Harvard University Press, 2001.

Antonello, Alessandro. "Protecting the Southern Ocean Ecosystem: The Environmental Protection Agenda of Antarctic Diplomacy and Science." In *International Organizations & Environmental Protection: Conservation and Globalization in the Twentieth Century,* edited by Wolfram Kaiser and Jan-Henrik Meyer, 268–92. New York: Berghahn Books, 2017.

Armiero, Marco, and Lise Sedrez, eds. *A History of Environmentalism: Local Struggles, Global Histories.* London: Bloomsbury Academic, 2014.

Aronova, Elena, Karen Baker, and Naomi Oreskes. "Big Science and Big Data in Biology: From the International Geophysical Year through the International Biological Program to the Long Term Ecological Research (LTER) Network, 1957–Present." *Historical Studies in the Natural Sciences* 40, no. 2 (2010): 183–224.

Avise, John C. *Conceptual Breakthroughs in Evolutionary Genetics: A Brief History of Shifting Paradigms.* London: Academic Press Elsevier, 2014.

AWG. "Working Group on the 'Anthropocene.'" Subcommission on Quaternary Stratigraphy, n.d. Accessed 23 March 2019. http://quaternary.stratigraphy.org/workinggroups/anthropocene/.

Ayres, Peter G. *Shaping Ecology: The Life of Arthur Tansley.* Oxford: John Wiley & Sons, 2012.

Bailey, Ronald. "Who Is Maurice Strong?" *National Review,* 1 September 1997, 32–32.

Baker, Susan. *Sustainable Development.* London: Routledge, 2015.

Barnes, Jessica. "Rifts or Bridges? Ruptures and Continuities in Human-Environment Interactions." In *Whose Anthropocene? Revisiting Dipesh Chakrabarty's "Four Theses,"* edited by Robert Emmett and Thomas Lekan. *RCC Perspectives: Transformations in Environment and Society* 2, 47–54. Munich: Rachel Carson Center, 2016.

Barrow, Mark V. *Nature's Ghosts: Confronting Extinction from the Age of Jefferson to the Age of Ecology*. Chicago: University of Chicago Press, 2009.

Bartlett, Robert V., Priya A. Kurian, and Madhu Malik. *International Organizations and Environmental Policy*. Westport: Greenwood Press, 1995.

Bashford, Alison. *Global Population: History, Geopolitics, and Life on Earth*. New York: Columbia University Press, 2014.

Batisse, Michel. "Of Mammoth and Men: Conservation for Human Survival." *UNESCO Courier* 33, no. 5 (1980): 4–8.

Beck, Silke. "Moving Beyond the Linear Model of Expertise? IPCC and the Test of Adaption." *Regional Environmental Change* 11, no. 2 (2011): 297–306.

Beltrán, Javier, ed. *Indigenous and Traditional Peoples and Protected Areas: Principles, Guidelines and Case Studies*. Cardiff: World Commission on Protected Areas (WCPA); IUCN; Cardiff University, 1998.

Benson, Etienne. "Territorial Claims: Experts, Antelopes, and the Biology of Land Use in Uganda, 1955–75." *Comparative Studies of South Asia, Africa and the Middle East* 35, no. 1 (2015): 137–55.

Berkes, Fikret. "Rethinking Community-Based Conservation." *Conservation Biology* 18, no. 3 (2004): 621–30.

Beumer, Carijn, and Pim Martens. "IUCN and Perspectives on Biodiversity Conservation in a Changing World." *Biodiversity and Conservation* 22, no. 13–14 (2013): 3105–20.

Bhouraskar, Digambar. *United Nations Development Aid: A Study in History and Politics*. New Delhi: Academic Foundation, 2007.

Biermann, Frank. "'Earth System Governance' as a Crosscutting Theme of Global Change Research." *Global Environmental Change* 17, no. 1 (2007): 326–37.

Bijker, Wiebe E., Roland Bal, and Ruud Hendriks. *The Paradox of Scientific Authority: The Role of Scientific Advice in Democracies*. Boston: MIT Press, 2009.

Blair, W. Frank. *Big Biology: The US/IBP*. Stroudsburg: Dowden, Hutchinson & Ross, 1977.

Bluwstein, Jevgeniy. "From Colonial Fortress to Neoliberal Landscape in Northern Tanzania: A Biopolitical Ecology of Wildlife Conservation." *Journal of Political Ecology* 25, no. 1 (2018): 144–68.

Boardman, Robert B. *International Organizations and the Conservation of Nature*. Bloomington: Indiana University Press, 1981.

Bocking, Stephen. "Conserving Nature and Building a Science: British Ecologists and the Origins of the Nature Conservancy." In *Science and Nature: Essays in the History of the Environmental Sciences*, vol. 8, edited by Michael Shortland, 89–114. Oxford: British Society for the History of Science, 1993.

———. *Ecologists and Environmental Politics: A History of Contemporary Ecology*. New Haven: Yale University Press, 1997.

———. *Nature's Experts: Science, Politics, and the Environment*. New Brunswick: Rutgers University Press, 2004.

Boote, Robert. "Obituary. Max Nicholson: The Prime Mover of the Nature Conservancy and the World Wildlife Fund, Who Helped Inspire Nature Reserves and Ecological Research." *The Guardian*. Published electronically 28 April 2003. Accessed 24 March 2019. http://www.theguardian.com/news/2003/apr/28/guardianobituaries.highereducation.

Borowy, Iris. *Defining Sustainable Development for Our Common Future: A History of the World Commission on Environment and Development (Brundtland Commission)*. London: Routledge, 2013.

Borstelmann, Thomas. *The 1970s: A New Global History from Civil Rights to Economic Inequality*. Princeton: Princeton University Press, 2012.

Bosc, Frédéric. *Tell Me about UNESCO*. Paris: UNESCO, 2001.

Bowler, Peter J., and Iwan Rhys Morus. *Making Modern Science: A Historical Survey*. Chicago: University of Chicago Press, 2005.

Brady, Lisa M. "Life in the DMZ: Turning a Diplomatic Failure into an Environmental Success." *Diplomatic History* 32, no. 4 (2008): 585–611.

Bramwell, Anna. *Ecology in the 20th Century: A History*. New Haven: Yale University Press, 1989.

Brockington, Dan. *Fortress Conservation: The Preservation of the Mkomazi Game Reserve, Tanzania*. Bloomington: Indiana University Press, 2002.

Büscher, Bram, Wolfram Heinz Dressler, and Robert Fletcher. *Nature Inc.: Environmental Conservation in the Neoliberal Age*. Tucson: University of Arizona Press, 2014.

Budowski, Gerardo. "The Biosphere to Come." In *The Environmental Future: Proceedings of the First International Conference on Environmental Future, Held in Finland from 27 June to 3 July 1971*, edited by Nicholas Polunin. London: Palgrave Macmillan, 1972.

———. "Should Ecology Conform to Politics?" *IUCN Bulletin*, n.s., 5, no. 12 (1974): 45–46.

Budowski, Gerardo, Frank Nicholls, and Raymond F. Dasmann. "Draft Programme and Budget for 1973–1975." In *Proceedings of the Eleventh General Assembly, Banff, Alberta, Canada 1972*, edited by IUCN. Morges: IUCN, 1972.

Buell, Lawrence. *The Environmental Imagination: Thoreau, Nature Writing, and the Formation of American Culture*. Cambridge, MA: Harvard University Press, 1996.

Caldwell, Lynton K. "Concepts in Development of International Environmental Policies." In *Proceedings of the Twelfth Technical Meeting, Banff, Alberta, Canada 1972*, edited by IUCN, 91–102. Morges: IUCN, 1972.

———. *In Defense of Earth: International Protection of the Biosphere*. Bloomington: Indiana University Press, 1972.

Caldwell, Lynton K., and Paul S. Weiland. *International Environmental Policy: From the Twentieth to the Twenty-First Century*. Durham, NC: Duke University Press, 1996.

Camprubí, Lino. "Review: The Invention of the Global Environment." *Historical Studies in the Natural Sciences* 42, no. 2 (2016): 243–51.

Carrington, Damian. "The Anthropocene Epoch: Scientists Declare Dawn of Human-Influenced Age." *The Guardian*, 29 August 2016. Accessed 23 March 2019. https://www.theguardian.com/environment/2016/aug/29/declare-anthropocene-epoch-experts-urge-geological-congress-human-impact-earth.

Carruthers, Jane. *National Park Science: A Century of Research in South Africa*. Cambridge: Cambridge University Press, 2017.

Carson, Rachel. *Silent Spring*. Boston: Houghton Mifflin Harcourt, 1962.

Castree, Noel. "The Anthropocene and the Environmental Humanities: Extending the Conversation." *Environmental Humanities* 5, no. 1 (2014): 233–60.

Chakrabarty, Dipesh. "The Climate of History: Four Theses." *Critical Inquiry* 35, no. 2 (2009): 197–222.

Chin, Anne. "Editorial Board." *Anthropocene* 1, no. 1 (2013): editorial.

Clapham, Arthur R., ed. *The IBP Survey of Conservation Sites: An Experimental Study*. Cambridge: Cambridge University Press, 1980.

Clarke, Sabine. "A Technocratic Imperial State? The Colonial Office and Scientific Research, 1940–1960." *Twentieth Century British History* 18, no. 4 (2007): 453–80.

Clements, Frederic E., and Victor E. Shelford. *Bio-Ecology*. London: Chapman & Hall, 1939.

Cochrane, Rexmond D. *The National Academy of Sciences: The First Hundred Years, 1863–1963*. Washington, DC: National Academies Press, 1978.

Coleman, David C. *Big Ecology: The Emergence of Ecosystem Science*. Berkeley: University of California Press, 2010.

Collins, Harry M. *Changing Order: Replication and Induction in Scientific Practice*. Chicago: University of Chicago Press, 1992.

Collins, Kenneth J., and Joseph S. Weiner. *Human Adaptability: A History and Compendium of Research in the International Biological Programme*. London: Taylor & Francis, 1977.

Commission on International Development, and Lester B. Pearson. *Partners in Development: Report*. New York: Praeger, 1969.

Committee of IUCN Members United Kingdom, and Council for Environmental Conservation United Kingdom. *Earth's Survival: A Conservation and Development Programme for the UK*. London: Programme Organizing Committee, UK Committee for IUCN, 1981–1982.

Commoner, Barry. *The Closing Circle: Man, Nature, and Technology*. New York: Knopf, 1971.

Connelly, Mathew J. *Fatal Misconception: The Struggle to Control World Population*. Cambridge, MA: Harvard University Press, 2008.

Coolidge, Harold J. "World Biosphere Conference: A Challenge to Mankind." *IUCN Bulletin*, n.s., 2, no. 9 (1968): 65–66.

Cooper, Gregory J. *The Science of the Struggle for Existence: On the Foundations of Ecology*. Cambridge: Cambridge University Press, 2007.

Corbera, Esteve, Laura Calvet-Mir, Hannah Hughes, and Matthew Paterson. "Patterns of Authorship in the IPCC Working Group III Report." *Nature Climate Change* 6, no. 1 (2016): 94–99.

Cox, C. Barry, and Peter D. Moore. *Biogeography: An Ecological and Evolutionary Approach*. London: John Wiley & Sons, 2010.

Craige, Betty Jean. *Eugene Odum: Ecosystem Ecologist and Environmentalist*. Athens, GA: University of Georgia Press, 2002.

Crawford, Elisabeth. *Nationalism and Internationalism in Science, 1880–1939: Four Studies of the Nobel Population*. Cambridge: Cambridge University Press, 2002.

Crofts, Roger, and Des Thompson. "Professor Duncan Poore––Global Conservationist." *Scottish Natural Heritage*, 12 April 2016. Accessed 24 March 2019. https://scotlands nature.wordpress.com/2016/04/12/professor-duncan-poore-global-conservationist/.

Cronon, William. *Uncommon Ground: Toward Reinventing Nature*. New York: W.W. Norton & Company, 1995.

Crutzen, Paul J., and Eugene F. Stoermer. "The Anthropocene." *Global Change Newsletter* 41 (2000): 17–18.

Cullather, Nick. *The Hungry World: America's Cold War Battle against Poverty in Asia*. Cambridge, MA: Harvard University Press, 2011.

Curtis, Adam. "The Use and Abuse of Vegetational Concepts." Episode 2 of *All Watched over by Machines of Loving Grace*. United Kingdom: BBC Two, 2011.

Dag Hammarskjöld Foundation. *What Now? The 1975 Dag Hammarskjöld Report: Prepared on the Occasion of the Seventh Special Session of the United Nations General Assembly.* Uppsala: Dag Hammarskjöld Foundation, 1975.

Dalby, Simon. "Framing the Anthropocene: The Good, the Bad and the Ugly." *Anthropocene Review* 3, no. 1 (2016): 33–51.

Dasmann, Raymond F. *African Game Ranching.* Oxford: Pergamon Press, 1964.

———. *Called by the Wild: The Autobiography of a Conservationist.* Berkeley: University of California Press, 2002.

———. *The Conservation Alternative.* New York: Wiley, 1975.

———. "Conservation and Rational Uses of the Environment." *Nature and Resources* 4 (1968): 2–5.

———. *Environmental Conservation.* New York: Wiley, 1965, 1968, 1972.

———. "Stockholm, Bucharest, and Rome—What Next?" *IUCN Bulletin*, n.s., 6, no. 1 (1975): 1.

———. *A System for Defining and Classifying Natural Regions for Purposes of Conservation: A Progress Report.* Morges: IUCN, 1973.

———. "Towards a System for Classifying Natural Regions of the World and Their Representation by National Parks and Reserves." *Biological Conservation* 4, no. 4 (1972): 247–55.

Dasmann, Raymond F., and Randall Jarrell. *Raymond F. Dasmann: A Life in Conservation Biology.* Bloomington: Xlibris Corporation, 2000.

Dasmann, Raymond F., John P. Milton, and Peter H. Freeman. *Ecological Principles for Economic Development.* New York: John Wiley & Sons, 1973.

Dasmann, Raymond F., and UNESCO. *Planet in Peril: Man and the Biosphere Today.* New York: World Pub., 1972.

Davenport, Coral. "With Trump in Charge, Climate Change References Purged from Website." *New York Times.* Published electronically 20 January 2017. https://nyti.ms/2jIq4RE.

De Bont, Raf. "Borderless Nature: Experts and the Internationalization of Nature Protection, 1890–1940." In *Scientists' Expertise as Performance*, edited by Joris Vandendriessche, Evert Peeters, and Kaat Wils. London: Pickering & Chatto, 2015.

———. "'Primitives' and Protected Areas: International Conservation and the 'Naturalization' of Indigenous People, ca. 1910–1975." *Journal of the History of Ideas* 76, no. 2 (2015): 215–36.

De Bont, Raf, Simone Schleper, and Hans Schouwenburg. "Conservation Conferences and Expert Networks in the Short 20th Century." *Environment and History* 23, no. 4 (2017): 569–99.

De Bont, Raf, and Geert Vanpaemel. "Editorial Introduction to Special Section, 'the Scientist as Activist: Biology and the Nature Protection Movement, 1900–1950.'" *Environment and History* 18, no. 2 (2012): 203–8.

De Pryck, Kari, and Krystel Wanneau. "(Anti-)Boundary Work in Global Environmental Change Research and Assessment." *Environmental Science and Policy* 77, no. 1 (2017): 203–10.

De Sadeleer, N. *Implementing the Precautionary Principle: Approaches from the Nordic Countries, EU and USA.* London: Earthscan, 2012.

Dean, Katrina, Simon Naylor, Simone Turchetti, and Martin Siegert. "Data in Antarctic Science and Politics." *Social Studies of Science* 38, no. 4 (2008): 571–604.

Deese, Richard S. "The New Ecology of Power: Julian and Aldous Huxley in the Cold War Era." In *Environmental Histories of the Cold War*, edited by J. R. McNeill and C. R. Unger, 279–300. Cambridge: Cambridge University Press, 2010.

Desai, Bharat H. "UNEP: A Global Environmental Authority." *Environmental Policy and Law* 36, no. 3–4 (2006): 137–57.

Di Castri, Francesco, Malcolm Hadley, and Jeanne Damlamian. "MAB: The Man and the Biosphere Program as an Evolving System." *Ambio* (1981): 52–57.

Diamond, Jared M. "The Island Dilemma: Lessons of Modern Biogeographic Studies for the Design of Natural Reserves." *Biological Conservation* 7, no. 2 (1975): 129–46.

Dienes, Margaret, and Robert Brown. *Diversity and Stability in Ecological Systems*. Springfield: Clearinghouse for Federal Scientific and Technical Information, 1969.

Disco, Cornelius, and Eda Kranakis, eds. *Cosmopolitan Commons: Sharing Resources and Risks across Borders*. Cambridge, MA: MIT Press, 2013.

Dorst, Jean. "A Biologist Looks at the Animal World (Beast and Men)." *UNESCO Courier* 22, no. 1 (1969): 17–22.

Dryzek, John S., and David Schlosberg. *Debating the Earth: The Environmental Politics Reader*. Oxford: Oxford University Press, 2005.

Dubos, René. *Man Adapting*. New Haven: Yale University Press, 1965.

Dunlap, Riley E., and Angela G. Mertig. *American Environmentalism: The U.S. Environmental Movement, 1970–1990*. New York: Taylor & Francis, 2014.

Dupuy, Pierre-Marie, and Jorge E. Viñuales. *International Environmental Law*. Cambridge, MA: Cambridge University Press, 2015.

Dworkin, Daniel M., and ICSU. *Environmental Sciences in Developing Countries: Summary Reports and Recommendations, Scope/UNEP Symposium on Environmental Sciences in Developing Countries, Nairobi, February 11–23, 1974*. Indianapolis: SCOPE Secretariat, 1974.

Edwards, Paul N. "Meteorology as Infrastructural Globalism." In *Global Power Knowledge: Science and Technology in International Affairs*, edited by John Krige and Kai-Henrik Barth, 229–50. Chicago: Chicago University Press, 2006.

———. *A Vast Machine: Computer Models, Climate Data, and the Politics of Global Warming*. Boston: MIT Press, 2010.

Egan, Michael. *Barry Commoner and the Science of Survival: The Remaking of American Environmentalism*. Cambridge, MA: MIT Press, 2007.

Ehrlich, Paul R. *The Population Bomb*. New York: Buccaneer Books, 1968.

Elichirigoity, Fernando. *Planet Management: Limits to Growth, Computer Simulation, and the Emergence of Global Spaces*. Chicago: Northwestern University Press, 1999.

Ellenberg, Heinz. *Vegetation Mitteleuropas mit den Alpen in kausaler, dynamischer und historischer Sicht*. Stuttgart: E. Ulmer, 1963.

Ellenberg, Heinz, and Dieter Mueller Dombois. *A Key to Raunkiaer Plant Life Forms with Revised Subdivisions*. Zurich: Stiftung Rübel, 1967.

Emmelin, Lars. "The Stockholm Conferences." *Ambio* 1, no. 4 (1972): 135–40.

Emmerij, Louis, Richard Jolly, and Thomas G. Weiss. *Ahead of the Curve? UN Ideas and Global Challenges*. Bloomington: Indiana University Press, 2001.

Engels, Jens Ivo. "Modern Environmentalism." In *The Turning Points of Environmental History*, edited by F. Uekötter, 119–31. Pittsburg: University of Pittsburg Press, 2010.

Evans, David. *A History of Nature Conservation in Britain.* London: Routledge, 1992.

FAO. *Report of the FAO Technical Conference on Marine Pollution and Its Effects on Living Resources and Fishing: Rome, 9–18 December 1970.* Rome: FAO, 1971.

———. *Shifting Cultivation and Soil Conservation in Africa.* Rome: FAO, 1974.

———. *The State of Food and Agriculture 2000.* Rome: FAO, 2000.

Farnham, Timothy. "A Confluence of Values: Historical Roots of Concern for Biological Diversity." In *The Routledge Handbook of Philosophy of Biodiversity*, edited by Justin Garson, Anya Plutynski, and Sahotra Sarkar, 11–25. London: Routledge, 2016.

Farvar, Mohammad Taghi, and John P. Milton, eds. *The Careless Technology: Ecology and International Development; The Record of the Conference on the Ecological Aspects of International Development, December 8–11, 1968, at Airlie House, Warrenton, Virginia.* Garden City, NY: Natural History Press, 1972.

———. *The Unforeseen International Ecologic Boomerang: Conference on the Ecological Aspects of International Development.* New York: American Museum of Natural History, 1969.

Ferguson, James. *The Anti-Politics Machine: "Development," Depoliticization, and Bureaucratic Power in Lesotho.* Minneapolis: University of Minnesota Press, 1990.

Fischer, Robert, Stewart Maginnis, William Jackson, Edmund Barrow, and Sally Jeanrenaud. *Linking Conservation and Poverty Reduction: Landscapes, People and Power.* Oxon: Earthscan, 2008.

Flippen, J. Brooks. "Richard Nixon, Russell Train, and the Birth of Modern American Environmental Diplomacy." *Diplomatic History* 32, no. 4 (2008): 613–38.

Forsdyke, A. G., and WMO. *Meteorological Factors in Air Pollution.* Geneva: Secretariat of the World Meteorological Organization, 1970.

Fosberg, Francis Raymond. "A Classification of Vegetation for General Purposes." *Tropical Ecology* 2 (1961): 1–28.

Franke, Anselm, Eyal Weizman, and Haus der Kulturen der Welt. *Forensis: The Architecture of Public Truth.* Berlin: Sternberg Press, 2014.

Freeman, Christopher. "Prometheus Unbound." *Futures* 16, no. 5 (1984): 494–507.

Frey, Marc, Sönke Kunkel, and Corinna R. Unger, eds. *International Organizations and Development, 1945–1990.* London: Palgrave Macmillan, 2014.

Fuller, R. Buckminster. *Operating Manual for Spaceship Earth.* Carbondale: Southern Illinois University Press, 1969.

Galison, Peter, and Bruce Hevly, eds. *Big Science: The Growth of Large-Scale Research.* Stanford: Stanford University Press, 1992.

Gandhi, Indira. *Man and His Environment.* New Delhi: India Book Centre, 1973.

Garavini, Giuliano. "Completing Decolonization: The 1973 'Oil Shock' and the Struggle for Economic Rights." *International History Review* 33, no. 3 (2011): 473–87.

Garavini, Giuliano, and Richard R. Nybakken. *After Empires: European Integration, Decolonization, and the Challenge from the Global South 1957–1986.* Oxford: Oxford University Press, 2012.

Gardner, Benjamin. *Selling the Serengeti: The Cultural Politics of Safari Tourism.* Athens, GA: Georgia University Press, 2016.

Gardner, Richard N. *Sterling Dollar Diplomacy: Anglo-American Collaboration in the Reconstruction of Multilateral Trade*. Oxford: Clarendon Press, 1956.

Garland, Elizabeth. "The Elephant in the Room: Confronting the Colonial Character of Wildlife Conservation in Africa." *African Studies Review* 51, no. 3 (2008): 51–74.

Gay, Hannah. *The Silwood Circle: A History of Ecology and the Making of Scientific Careers in Late Twentieth-Century Britain*. London: Imperial College Press, 2013.

"General Assembly Declaration on the Establishment of a New International Economic Order." *American Journal of International Law* 68, no. 4 (1974): 798–801.

Ghabbour, Samir I. "Scope/UNEP Symposium: Environmental Sciences in Developing Countries, Held in the Kenyatta Conference Centre, Nairobi, Kenya, 11–23 February 1974." *Environmental Conservation* 1, no. 3 (1974).

Gibson, Mark. *The Feeding of Nations: Redefining Food Security for the 21st Century*. New York: CRC Press, 2016.

Giere, Ronald N. "Controversies Involving Science and Technology: A Theoretical Perspective." In *Scientific Controversies: Case Studies in the Resolution and Closure of Dispute in Science and Technology*, edited by Hugo Tristram Egelhardt and Arthur L. Caplan, 125–50. Cambridge: Cambridge University Press, 1987.

Gieryn, Thomas F. "Boundary-Work and the Demarcation of Science from Non-Science: Strains and Interests in Professional Ideologies of Scientists." *American Sociological Review* 48, no. 6 (1983): 781–95.

———. *Cultural Boundaries of Science: Credibility on the Line*. Chicago: University of Chicago Press, 1999.

Gillette, Robert. "The Limits to Growth: Hard Sell for a Computer View of Doomsday." *Science* 175, no. 4026 (1972): 1088–92.

Gilman, Nils. "The New International Economic Order: A Reintroduction." *Humanity: An International Journal of Human Rights, Humanitarianism, and Development* 6, no. 1 (2015): 1–16.

Gissibl, Bernhard. "Die Mythen der Serengeti: Naturbilder, Naturpolitik und die Ambivalenz westlicher Um-Weltbürgerschaft in Ostafrika." *Denkanstöße. Stiftung Natur und Umwelt Rheinland-Pfalz* 10 (2013): 48–76.

———. *The Nature of German Imperialism: Conservation and the Politics of Wildlife in Colonial East Africa*. New York: Berghahn Books, 2016.

Gissibl, Bernhard, Sabine Höhler, and Patrick Kupper. *Civilizing Nature: National Parks in Global Historical Perspective*. New York: Berghahn Books, 2012.

Glowka, Lyle, Françoise Burhenne-Guilmin, Hugh Synge, Jeffrey McNeely, and Lothar Gründling. *A Guide to the Convention on Biological Diversity*. Gland: IUCN, 1994.

Goldsmith, Edward, and Robert Allen. *Blueprint for Survival*. New York: New American Library, 1972.

Goldsmith, Frank B. "An Assessment of the Fosberg and Ellenberg Methods of Classifying Vegetation for Conservation Purposes." *Biological Conservation* 6, no. 1 (1974): 3–6.

Golley, Frank B. *A History of the Ecosystem Concept in Ecology: More Than the Sum of the Parts*. New Haven: Yale University Press, 1993.

Gowdy Wygant, Cecilia. "The United Nations Conference on the Human Environment: Formation, Significance and Political Challenges." Master's thesis, Texas Tech University, 2004.

Graham, Edward. "A New Network of Reserves." *New Scientist*, 14 October 1965.

Graham, Otis L., ed. *Environmental Politics and Policy, 1960s–1990s*. University Park: Pennsylvania State University Press, 2010.

Grebenik, Eugene. "World Population and Resources." *Political Quarterly* 26, no. 4 (1955): 371–79.

Greenaway, Frank. *Science International: A History of the International Council of Scientific Unions*. Cambridge: Cambridge University Press, 2006.

Greenough, Paul R., and Anna L. Tsing. *Nature in the Global South: Environmental Projects in South and Southeast Asia*. Durham, NC: Duke University Press, 2003.

Greschke, Heike M., and Julia Tischler. *Grounding Global Climate Change: Contributions from the Social and Cultural Sciences*. Dordrecht: Springer Netherlands, 2014.

Griffin, Andrew. "Scientists to Oppose Donald Trump in Huge 'March for Science' in Washington." *The Independent*, 27 January 2017. Accessed 3 March 2018. http://www.independent.co.uk/news/science/donald-trump-science-march-washington-climate-change-global-warming-a7547206.html.

Grove, Richard H. *Green Imperialism: Colonial Expansion, Tropical Island Edens and the Origins of Environmentalism, 1600–1860*. Cambridge: Cambridge University Press, 1995.

Grundmann, Reiner. "The Problem of Expertise in Knowledge Societies." *Minerva* 55, no. 1 (2017): 25–48.

Gwynne, Michael D. "The Global Environment Monitoring System (GEMS) of UNEP." *Environmental Conservation* 9, no. 1 (1982): 35–41.

Haas, Ernst B., and John G. Ruggie. "What Message in the Medium of Information Systems?" *International Studies Quarterly* 26, no. 2 (1982): 190–219.

Haas, Peter M. "Banning Chlorofluorocarbons: Epistemic Community Efforts to Protect Stratospheric Ozone." *International Organizations* 46, no. 1 (1992): 187–224.

———. "Constructing Environmental Conflicts from Resource Scarcity." *Global Environmental Politics* 2, no. 1 (2002): 1–11.

———. *Saving the Mediterranean: The Politics of International Environmental Cooperation*. New York: Columbia University Press, 1990.

Hagen, Joel B. *An Entangled Bank: The Origins of Ecosystem Ecology*. New Brunswick: Rutgers University Press, 1992.

———. "Teaching Ecology during the Environmental Age, 1965–1980." *Environmental History* 13, no. 4 (2008): 704–23.

Hallonsten, Olof. *Big Science Transformed: Science, Politics and Organization in Europe and the United States*. London: Palgrave Macmillan, 2016.

Hamblin, Jacob D. "Environmental Diplomacy in the Cold War: The Disposal of Radioactive Waste at Sea During the 1960s." *International History Review* 24, no. 2 (2002): 348–75.

———. "Gods and Devils in the Details: Marine Pollution, Radioactive Waste, and an Environmental Regime circa 1972." *Diplomatic History* 32, no. 4 (2008): 539–60.

Hamilton, Clive. "Human Destiny in the Anthropocene." In *The Anthropocene and the Global Environmental Crisis: Rethinking Modernity in a New Epoch*, edited by Clive Hamilton, François Gemenne, and Christophe Bonneuil. Abington-on-Thames: Earthscan, 2015.

Hammond, Debora. *The Science of Synthesis: Exploring the Social Implications of General Systems Theory*. Boulder: University Press of Colorado, 2003.

Haraway, Donna J. "In the Beginning Was the Word: The Genesis of Biological Theory." *Signs* 6, no. 3 (1981): 469–81.

Hardin, Garrett. *The Tragedy of the Commons*. Washington, DC: American Association for the Advancement of Science, 1968.

Hays, Cassie M. "Placing Nature(s) on Safari." *Tourist Studies* 12, no. 3 (2012): 250–67.

Heise, Ursula K. *Sense of Place and Sense of Planet: The Environmental Imagination of the Global*. New York: Oxford University Press, 2008.

Hilgartner, Stephen. *Science on Stage: Expert Advice as Public Drama*. Stanford: Stanford University Press, 2000.

Hilton, Matthew. *Consumerism in Twentieth-Century Britain: The Search for a Historical Movement*. Cambridge: Cambridge University Press, 2003.

Hixson, Walter. *American Foreign Relations: A New Diplomatic History*. New York: Taylor & Francis, 2015.

Höhler, Sabine. *Spaceship Earth in the Environmental Age, 1960–1990*. London: Pickering & Chatto, 2015.

———. "Von Biodiversität zu Biodiversifizierung: Eine neue Ökonomie der Natur?" *Berichte zur Wissenschaftsgeschichte* 37, no. 1 (2014): 60–77.

Holdgate, Martin W. *The Green Web: A Union for World Conservation*. Gland: IUCN, 1999.

Holdgate, Martin W., Mohamed Kassas, Gilbert F. White, and UNEP. *The World Environment 1972–1982: A Report*. Dublin: Tycooly International Pub., 1982.

Holling, Crawford S. "Resilience and Stability of Ecological Systems." *Annual Review of Ecology and Systematics* 4 (1973): 1–23.

Holling, Crawford S., and UNEP. *Adaptive Environmental Assessment and Management*. Laxenburg: International Institute for Applied Systems Analysis, 1978.

Horberry, John. *Status and Application of Environmental Impact Assessment for Development*. Nairobi: Conservation for Development Centre, 1984.

Htun, M. Nay. "Development of UNEP Guidelines for Assessing Industrial Environmental Impact and Environmental Criteria for the Siting of Industry." In *Perspectives on Environmental Impact Assessment*, edited by Brian D. Clark, Alexander Gilad, Ronald Bisset, and Paul Tomlinson, 253–63. Dordrecht: Springer Netherlands, 1984.

Hughes, Agatha C., and Thomas P. Hughes. *Systems, Experts, and Computers: The Systems Approach in Management and Engineering, World War II and After*. Boston: MIT Press, 2011.

Hughes-Evans, David. "Interview: Profile of Harold Jefferson Coolidge." *Environmentalist* 1, no. 1 (1981): 65–74.

Hünemörder, Kai F. *Die Frühgeschichte der globalen Umweltkrise und die Formierung der deutschen Umweltpolitik* (1950–1973). Stuttgart: Franz Steiner Verlag, 2004.

———. "Environmental Crisis and Soft Politics: Détente and the Global Environment, 1968–1975." In *Environmental Histories of the Cold War*, edited by Daniel McNeill and Corinna R. Unger, 257–78. Cambridge: Cambridge University Press, 2010.

Hutchinson, G. Evelyn, Abraham H. Oort, George M. Woodwell, H. L. Penman, Bert Bolin, Preston Cloud, Aharon Gibor, C. C. Delwich, Edward S. Deevey Jr., Lester R. Brown, S. Fred Singer, and Harrison Brown. *The Biosphere*. Scientific American book series. New York: W. H. Freeman, 1970.

Huxley, Julian. *Evolution: The Modern Synthesis*. New York: Harper & Brothers, 1943.

———. *The Stream of Life*. London: Watts, 1926.

Huxley, Julian, H. Levy, T. D. Barlow, and P. M. S. Blackett. *Scientific Research and Social Needs*. London: Watts & Co., 1934.

Igoe, Jim. *The Nature of Spectacle: On Images, Money and Conservation Capitalism*. Tucson: University of Arizona Press, 2017.

IIED. "Remembering Duncan Poore, a Leading Light in Sustainable Forestry." IIED website, 19 October 2016. Accessed 24 March 2019. https://www.iied.org/remembering-duncan-poore-leading-light-sustainable-forestry.

Immerwahr, Daniel. *Thinking Small: The United States and the Lure of Community Development*. Cambridge, MA: Harvard University Press, 2015.

Isenberg, Andrew C. *The Oxford Handbook of Environmental History*. Oxford: Oxford University Press, 2014.

Ishwaran, Natarajan, Ana Persic, and Nguyen H. Tri. "Concept and Practice: The Case of UNESCO Biosphere Reserves." *International Journal of Environment and Sustainable Development* 7, no. 2 (2008): 118–31.

IUCN. *Draft, International Covenant on Environment and Development – Implementing Sustainability – Fifth Edition: Updated Text*. Gland: IUCN, 2015.

———. "International Day for Biological Diversity 2015: Biodiversity for Sustainable Development." IUCN website, 22 May 2015. Accessed 23 March 2019. https://www.iucn.org/content/international-day-biological-diversity-2015-biodiversity-sustainable-development.

———. "IUCN Prepares World Strategy." *IUCN Bulletin*, n.s., 8, no. 10 (1977): 59.

———. *IUCN Programme 2017–2020: Approved by the IUCN World Conservation Congress, September 2016*. Gland: IUCN, 2016.

———. "IUCN's Programme 1976–1978." *IUCN Bulletin*, n.s., 6, no. 11 (1975): 43–45.

———. *IUCN Yearbook: Annual Report*. Morges: IUCN, 1972, 1975.

———. "Members." IUCN website, n.d. Accessed 15 March 2019. https://www.iucn.org/about/union/members.

———, ed. *Proceedings of a Regional Meeting on the Use of Ecological Guidelines for Development in the Tropical Forest Areas of South East Asia, Held at Bandung, Indonesia 29 May to 1 June 1974*. Morges: IUCN, 1975.

———, ed. *Proceedings of the Eighth General Assembly, Kenya 1963*. Morges: IUCN, 1964.

———, ed. *Proceedings of the Eleventh General Assembly, Banff, Alberta 1972*. Morges: IUCN, 1972.

———, ed. *Proceedings of the Eleventh Technical Meeting, New Delhi 1969*. Morges: IUCN, 1969.

———, ed. *Proceedings of the Fifteenth General Assembly of IUCN, Christchurch 1981*. Gland: IUCN, 1983.

———, ed. *Proceedings of the Fifth General Assembly: Edinburgh, June 1956*. London: Society for the Promotion of Nature Reserves, in collaboration with the Nature Conservancy for the IUCN, 1957.

———, ed. *Proceedings of the Fourteenth General Assembly, Ashkhabad 1978*. Morges: IUCN, 1979.

———, ed. *Proceedings of the International Meeting on the Use of Ecological Guidelines for Development in the American Humid Tropics, Held at Caracas, Venezuela, 20–22 February 1974*. Morges: IUCN, 1975.

———, ed. *Proceedings of the Members' Assembly: World Conservation Congress Honolulu, Hawai'i, United States of America, 6–10 September 2016*. Gland: IUCN, 2016.

———, ed. *Proceedings of the Ninth General Assembly, Lucerne 1966*. Morges: IUCN, 1966.

———, ed. *Proceedings of the Seventeenth Session of the General Assembly of IUCN and Seventeenth IUCN Technical Meeting: San José, Costa Rica 1–10 February 1988*. Gland: IUCN, 1988.

———, ed. *Proceedings of the Seventh General Assembly, Warsaw 1960*. Morges: IUCN, 1960.

———, ed. *Proceedings of the Tenth General Assembly, New Delhi 1969*. Morges: IUCN, 1969.

———, ed. *Proceedings of the Thirteenth (Extraordinary) General Assembly, Geneva 1977*. Morges: IUCN, 1977.

———, ed. *Proceedings of the Twelfth General Assembly, Kinshasa 1975*. Morges: IUCN, 1976.

———, ed. *Proceedings of the Twelfth Technical Meeting, Banff, Alberta, Canada 1972*. Morges: IUCN, 1972.

———. *Results of the IUCN Programme 2005–2008*. Gland: IUCN, 2009.

———. "UNEP-IUCN Link in Ecosystem Conservation." *IUCN Bulletin*, n.s., 6, no. 6 (1975): 23.

———. "The Union." IUCN website. https://www.iucn.org/secretariat/about/union.

———. "The World Conservation Strategy at a Glance." *IUCN Bulletin*, n.s., 3 (1980): 38–39.

IUCN, FAO, and UNESCO. *CCTA/IUCN Symposium on the Conservation of Nature and Natural Resources in Modern African States (in Collaboration with the FAO and UNESCO), Arusha, Tanganyika, 5–12 September 1961*. Lagos: Commission for Technical Cooperation in Africa South of the Sahara, 1962.

IUCN, and ICNP. *United Nations List of National Parks and Equivalent Reserves*. Morges: IUCN International Commission on National Parks, 1974.

IUCN, and UNESCO. *The Biosphere Reserve and Its Relationship to Other Protected Areas*. Gland: IUCN, 1979.

IUCN, United States Park Service, and Hugh F. I. Elliott, eds. *Second World Conference on National Parks: Yellowstone and Grand Teton National Parks*. Morges: IUCN, 1972.

IUCN, WWF, and UNEP. World Conservation Strategy. Gland: IUCN, 1980.

IUPN. *Proceedings and Reports of the Second Session of the General Assembly, Brussels 1950*. Brussels: IUPN, 1951.

Jamison, Andrew. "National Political Cultures and the Exchange of Knowledge: The Case of Systems Ecology." In *Denationalizing Science: The Contexts of International Scientific Practice*, edited by Elisabeth Crawford, Terry Shinn, and Sverker Sörlin, 187–208. Dordrecht: Springer Science and Business Media, 2013.

Jasanoff, Sheila. *The Fifth Branch: Science Advisers as Policymakers*. Cambridge, MA: Harvard University Press, 1990.

———. "Genealogies of STS." *Social Studies of Science* 42, no. 3 (2002): 435–41.

———. "A New Climate for Society." *Theory, Culture & Society* 27, no. 2–3 (2010): 223–53.

———. *Science at the Bar: Law, Science, and Technology in America*. Cambridge, MA: Harvard University Press, 2009.

———. "Speaking Honestly to Power. Review of *The Honest Broker: Making Sense of Science in Policy and Politics*, by Roger A. Pielke." *American Scientist* 96, no. 3 (2008): 240–43.

———, ed. *States of Knowledge: The Co-Production of Science and Social Order*. New York: Routledge, 2004.

Johnson, Brian D., and WWF-UK. *The Conservation and Development Programme for the U.K.: A Response to the World Conservation Strategy. An Overview--Resourceful Britain*. London: Kogan Page, 1983.

Johnson, Stanley. UNEP. *The First 40 Years: A Narrative*. Nairobi: UNON/Publishing Section Service, 2012.

Jørgensen, Dolly, Finn Arne Jørgensen, and Sara B. Pritchard. *New Natures: Joining Environmental History with Science and Technology Studies*. Pittsburg: University of Pittsburgh Press, 2013.

Kaiser, David, and W. Patrick McCray. *Groovy Science: Knowledge, Innovation, and American Counterculture*. Chicago: University of Chicago Press, 2016.

Kaiser, Wolfram, and Jan-Henrik Meyer, eds. *International Organizations & Environmental Protection: Conservation and Globalization in the Twentieth Century*. New York: Berghahn Books, 2017.

———. "Introduction: International Organizations and Environmental Protection in the Global Twentieth Century." In *International Organizations & Environmental Protection: Conservation and Globalization in the Twentieth Century*, edited by Wolfram Kaiser and Jan-Henrik Meyer, 1–30. New York: Berghahn Books, 2017.

Kassas, Mohamed. "Address at the Opening of the 15th Session of the IUCN General Assembly." In *Proceedings of the Fifteenth General Assembly of IUCN, Christchurch 1981*, edited by IUCN, 83–86. Gland: IUCN, 1981.

Keats, Jonathon. *You Belong to the Universe: Buckminster Fuller and the Future*. New York: Oxford University Press, 2016.

Keller, David R., and Frank B. Golley. *The Philosophy of Ecology: From Science to Synthesis*. Athens, GA: University of Georgia Press, 2000.

Ki-moon, Ban. *The Road to Dignity by 2030: Ending Poverty, Transforming All Lives and Protecting the Planet; Synthesis Report of the Secretary-General on the Post-2015 Agenda*. New York: UN, 2014.

Kingsland, Sharon E. "Designing Nature Reserves: Adapting Ecology to Real-World Problems." *Endeavour* 26, no. 1 (2002): 9–14.

———. *The Evolution of American Ecology, 1890–2000*. Baltimore: Johns Hopkins Press, 2005.

Klein, Naomi. *This Changes Everything: Capitalism vs. the Climate*. New York: Simon & Schuster, 2014.

Kline, Ronald R. *The Cybernetics Moment: Or Why We Call Our Age the Information Age*. Baltimore: Johns Hopkins University Press, 2015.

Kolbert, Elizabeth. *The Sixth Extinction: An Unnatural History*. New York: Henry Holt and Company, 2014.

Kozymka, Irena. *The Diplomacy of Culture: The Role of UNESCO in Sustaining Cultural Diversity*. New York: Palgrave Macmillan, 2014.

Krementsov, Nikolai. *International Science between the World Wars: The Case of Genetics*. New York: Routledge, 2004.

Krige, John. *American Hegemony and the Postwar Reconstruction of Science in Europe*. Cambridge, MA: MIT Press, 2006.

Krige, John, and Kai-Henrik Barth, eds. *Global Power Knowledge: Science and Technology in International Affairs*. Chicago: University of Chicago Press, 2006.

———. "Science, Technology, and International Affairs: New Perspectives." In *Global Power Knowledge: Science and Technology in International Affairs*, edited by John Krige and Kai-Henrik Barth, 1–24. Chicago: Chicago University Press, 2006.

Küchler, August W., Jorge M. Montoya Maquin, and UNESCO. *The UNESCO Classification of Vegetation: Some Tests in the Tropics*. Paris: UNESCO, 1970.

Kumi, Emmanuel, Albert A. Arhin, and Thomas Yeboah. "Can Post-2015 Sustainable Development Goals Survive Neoliberalism? A Critical Examination of the Sustainable Development–Neoliberalism Nexus in Development Countries." *Environment, Development and Sustainability* 16, no. 3 (2014): 539–54.

Kunkel, Sönke. "Contesting Globalization: The United Nations Conference on Trade and Development and the Transnationalization of Sovereignty." In *International Organizations and Development, 1945–1990*, edited by Marc Frey, Sönke Kunkel, and Corinna R. Unger, 240–58. London: Palgrave Macmillan, 2014.

Kupper, Patrick. *Creating Wilderness: A Transnational History of the Swiss National Park*. New York: Berghahn Books, 2014.

———. "Die '1970er Diagnose.' Grundsätzliche Überlegungen zu einem Wendepunkt der Umweltgeschichte." *Archiv für Sozialgeschichte* 43 (2003): 325–48.

Kwa, Chunglin. "Representations of Nature Mediating between Ecology and Science Policy: The Case of the International Biological Programme." *Social Studies of Science* 17, no. 3 (1987): 413–42.

Lamb, Robert. *Promising the Earth*. London: Routledge, 2012.

Lantier, Patricia. *Rachel Carson: Fighting Pesticides and Other Chemical Pollutants*. St. Catharines: Crabtree Publishing Company, 2009.

Latour, Bruno. *Down to Earth: Politics in the New Climatic Regime*. Cambridge & Medford: Polity Press, 2018.

———. *Politics of Nature: How to Bring the Sciences into Democracy*. Cambridge, MA: Harvard University Press, 2004.

———. *We Have Never Been Modern*. Cambridge, MA: Harvard University Press, 1993.

Laubichler, Manfred D., and Jürgen Renn. "Extended Evolution: A Conceptual Framework for Integrating Regulatory Networks and Niche Construction." *Journal of Experimental Zoology Part B: Molecular and Developmental Evolution* 324, no. 7 (2015): 565–77.

Leopold, Aldo. *A Sand County Almanac, and Sketches Here and There*. Oxford: Oxford University Press, 1949.

Lettevall, Rebecka, Geert J. Somsen, and Sven Widmalm, eds. *Neutrality in Twentieth-Century Europe: Intersections of Science, Culture, and Politics after the First World War*. London: Routledge, 2012.

Lewis, Michael L., ed. *American Wilderness: A New History*. Oxford: Oxford University Press, 2007.

Lockwood, Michael, Graeme Worboys, and Ashish Kothari. *Managing Protected Areas: A Global Guide*. London: Earthscan, 2012.

Loftas, Tony. "The UN's Agents of Change." *New Scientist* 66, no. 953 (1975): 608–9.

Lomolino, Mark V., Dov F. Sax, and James H. Brown. *Foundations of Biogeography: Classic Papers with Commentaries*. Chicago: University of Chicago Press, 2004.

Longino, Helen E. *Science as Social Knowledge: Values and Objectivity in Scientific Inquiry*. Princeton: Princeton University Press, 1990.

Long Martello, Marybeth, and Sheila Jasanoff. "Introduction: Globalization and Environmental Governance." In *Earthly Politics: Local and Global in Environmental Governance*, edited by Marybeth Long Martello and Sheila Jasanoff, 1–30. Cambridge, MA: MIT Press, 2004.

Long Martello, Marybeth, and Sheila Jasanoff, eds. *Earthly Politics: Local and Global in Environmental Governance*. Cambridge, MA: MIT Press, 2004.

Luke, Timothy W. "On Environmentality: Geo-Power and Eco-Knowledge in the Discourses of Contemporary Environmentalism." *Cultural Critique*, no. 31 (1995): 57–81.

MacArthur, Robert H., and Edward O. Wilson. "An Equilibrium Theory of Insular Zoogeography." *Evolution* (1963): 373–87.

———. *The Theory of Island Biogeography*. Princeton: Princeton University Press, 1967.

Macekura, Stephen. *Of Limits and Growth: The Rise of Global Sustainable Development in the Twentieth Century*. Cambridge: Cambridge University Press, 2015.

Mahiou, Ahmend. "Declaration on the Establishment of a New International Economic Order." United Nations Audiovisual Library of International Law, 2011. Accessed 23 March 2019. http://legal.un.org/avl/pdf/ha/ga_3201/ga_3201_e.pdf.

March for Science. "AAAS Advocacy Toolkit," n.d. Accessed 23 March 2019. https://aas.org/advocacy-resources/reference-guide-how-advocate-science.

Marsh, George P. *Man and Nature*. Cambridge, MA: Belknap Press, 1864.

Maslin, Mark A., and Simon L. Lewis. "Anthropocene: Earth System, Geological, Philosophical and Political Paradigm Shifts." *Anthropocene Review* 2, no. 2 (2015): 1–9.

Mauch, Christof. *Das Neue Rachel Carson Center in München Oder Was heisst und zu welchem Ende betreibt man Weltumweltgeschichte?* RCC Perspectives: Transformations in Environment and Society. Munich: Rachel Carson Center, 2010.

Maurel, Chloé. "L'UNESCO, un Pionnier de l'Ecologie?" *Monde(s)* 1, no. 3 (2013): 171–92.

Mayr, Ernst. *Systematics and the Origin of Species, from the Viewpoint of a Zoologist*. New York: Columbia University Press, 1942.

M'Bow, Amadou-Mahtar. "Man and the Biosphere." *UNESCO Courier* 34, no. 4 (1981): 4–5.

McCall Skutsch, Margaret, and Robin T. Flowerdew. "Measurement Techniques in Environmental Impact Assessment." *Environmental Conservation* 3, no. 3 (1976): 209–17.

McClellan, James E., and Harold Dorn. *Science and Technology in World History: An Introduction*. Baltimore: Johns Hopkins University Press, 2008.

McCormick, John. "The Origins of the World Conservation Strategy." *Environmental Review* 10, no. 3 (1986): 177–87.

———. *Reclaiming Paradise: The Global Environmental Movement*. Bloomington: Indiana University Press, 1991.

McDougall, Walter A. "Technocracy and Statecraft in the Space Age: Toward the History of a Saltation." *American Historical Review* 87, no. 4 (1982): 1010–40.

McHale, John. *The Ecological Context*. New York: G. Braziller, 1970.

———. *The Future of the Future*. New York: G. Braziller, 1969.

McIntosh, Robert. *The Background of Ecology: Concept and Theory*. Cambridge: Cambridge University Press, 1986.

McNeely, Jeffrey, Kenton Miller, Russell Mittermeier, Walter Reid, and Timothy Werner. *Conserving the World's Biological Diversity*. Gland; Washington, DC: IUCN, WRI, WWF-US, World Bank, 1990.

McNeill, John R., and Peter Engelke. *The Great Acceleration: An Environmental History of the Anthropocene since 1945*. Cambridge, MA: The Belknap Press of Harvard University Press, 2014.

McNeill, John R., and Erin S. Mauldin. *A Companion to Global Environmental History*. Chichester: John Wiley & Sons, 2014.

McNeill, John R., and Corinna R. Unger, eds. *Environmental Histories of the Cold War*. Cambridge: Cambridge University Press, 2010.

Mead, Margaret. "Anthropology and Ekistics." *Ekistics* 21, no. 123 (1966): 88–89.

Meadows, Donella H., Dennis L. Meadows, Jorgen Randers, and William W. Behrens. *The Limits to Growth: A Report for the Club of Rome's Project on the Predicament of Mankind*. New York: Universe Books, 1972.

Meine, Curt D., and Wendell Berry. *Aldo Leopold: His Life and Work*. Madison: University of Wisconsin Press, 2010.

Meisler, Stanley. *United Nations: A History*. New York: Grove Press, 2011.

Merchant, Carolyn. *American Environmental History: An Introduction*. New York: Columbia University Press, 2007.

Meyer, Jan-Henrik. "Sammelrezension [Reviews]: Where Did Environmentalism Come From? Rome, Adam: *The Genius of Earth Day. How a 1970 Teach-in Unexpectedly Made the First Green Generation*. New York 2013 / Hamblin, Jacob Darwin: *Arming Mother Nature. The Birth of Catastrophic Environmentalism*. New York 2013 / Zelko, Frank: *Make It a Green Peace! The Rise of a Countercultural Environmentalism*. New York 2013." *H-Soz-Kult*, 21 July 2016. http://www.hsozkult.de/publicationreview/id/rezbuecher-22483.

Meyer, John W., David J. Frank, Ann Hironaka, Evan Schofer, and Nancy B. Tuma. "The Structuring of a World Environmental Regime, 1870–1990." *International Organization* 51, no. 4 (1997): 623–51.

M'Gonigle, R. Michael, and Mark W. Zacher. *Pollution, Politics, and International Law: Tankers at Sea*. Berkeley: University of California Press, 1981.

Miller, Clark A. "Climate Science and the Making of a Global Political Order." In *States of Knowledge: The Co-Production of Science and Social Order*, edited by Sheila Jasanoff, 46–66. New York: Routledge, 2004.

Miller, Clark A., and Paul N. Edwards. *Changing the Atmosphere: Expert Knowledge and Environmental Governance*. Cambridge, MA: MIT Press, 2001.

Miller, Robert R., and IUCN. *Red Data Book*. Morges: IUCN, 1977.

Milman, Oliver. "Donald Trump Picks Climate Change Skeptic Scott Pruitt to Lead EPA." *The Guardian*, 8 December 2016. Accessed 24 March 2019. https://www.theguardian.com/us-news/2016/dec/07/trump-scott-pruitt-environmental-protection-agency.

Mitchell, Tim. *Rule of Experts: Egypt, Techno-Politics, Modernity*. Berkeley: University of California Press, 2002.

Morgan, Edward A., and Natalie Osborne. "It's the Lungfish, Stupid: Knowledge Fights, Activism, and the Science-Policy Interface." *Geographical Research* 54, no. 4 (2016): 365–76.

Muehlebach, Andrea. "'Making Place' at the United Nations: Indigenous Cultural Politics at the UN Working Group on Indigenous Populations." *Cultural Anthropology* 16, no. 3 (2001): 415–48.

Mueller-Dombois, Dieter. *Classification and Mapping of Plant Communities: A Review with Emphasis on Tropical Vegetation*. Chichester: John Wiley and Sons, 1984.

Mulligan, Martin, and Stuart Hill. *Ecological Pioneers: A Social History of Australian Ecological Thought and Action*. Cambridge: Cambridge University Press, 2001.

Mumford, Lewis. *Technics and Civilization*. New York: Harcourt, Brace, 1934.

Munn, Robert E., ICSU, and SCOPE. *Environmental Impact Assessment: Principles and Procedures*. Toronto: SCOPE, 1975.

Munro, David A. "Conservation and Politics." *IUCN Bulletin*, n.s., 10, no. 1 (1979): 1.

———. "Towards a World Strategy of Conservation." *Environmental Conservation* 6, no. 3 (1979): 169–70.

Munro, David A., IUCN, UNEP, and WWF. *Caring for the Earth: A Strategy for Sustainable Living*. Gland: IUCN, 1991.

Murphy, Craig N. *The United Nations Development Programme: A Better Way?* Cambridge: Cambridge University Press, 2006.

Nash, Roderick. *Wilderness and the American Mind*. New Haven: Yale University Press, 1965.

Needell, Allan A. "From Military Research to Big Science: Lloyd Berkner and Science-Statesmanship in the Postwar Era." In *Big Science: The Growth of Large-Scale Research*, edited by Peter Galison and Bruce W. Hevly, 290–311. Stanford: Stanford University Press, 1992.

Nelkin, Dorothy. *Controversy: Politics of Technical Decisions*. New York: Sage Publications, 1979.

———. *Nuclear Power and Its Critics: The Cayuga Lake Controversy*. New York: Cornell University Press, 1971.

Nerfin, Marc, and Fernando H. Cardoso. *Another Development: Approaches and Strategies*. Uppsala: Dag Hammarskjöld Foundation, 1977.

Neumann, Roderick P. *Imposing Wilderness: Struggles over Livelihood and Nature Preservation in Africa*. Berkeley: University of California Press, 1998.

———. "The Postwar Conservation Boom in British Colonial Africa." *Environmental History* 7, no. 1 (2002): 22–47.

———. "Ways of Seeing Africa: Colonial Recasting of African Society and Landscape in Serengeti National Park." *Ecumene* 2, no. 2 (1995): 149–69.

Nicholson, E. Max. *The Big Change: After the Environmental Revolution*. New York: McGraw-Hill, 1973.

———. *Britain's Nature Reserves*. London: Country Life, 1957.

———. *Conservation and the Next Renaissance: The Horace M. Albright Conservation Lectureship*. Berkeley: University of California, Berkeley, School of Forestry, 1964.

———. *The Environmental Revolution: A Guide for the New Masters of the World*. London: Penguin Books, 1970.

———. *Handbook to the Conservation Section of the International Biological Programme*. London: IBP/CT, 1968.

———. *How Britain's Resources Are Mobilized*. Oxford: Clarendon Press, 1940.

———. "International Economic Development and the Environment." *Journal of International Affairs* 24, no. 2 (1970): 272–87.

———. "Research and Natural Areas." In *First World Conference on National Parks: Proceedings of a Conference, Seattle, Washington, June 30–July 7, 1962*, edited by Alexander B.

Adams, IUCN, and UNESCO. Washington, DC: National Park Service, U.S. Department of the Interior, 1964.

Nnadozie, Kent. *African Perspectives on Genetic Resources: A Handbook on Laws, Policies, and Institutions Governing Access and Benefit-Sharing*. Washington, DC: Environmental Law Institute, 2003.

Odum, Eugene P. *Ecology*. New York: Holt, Rinehart and Winston, 1963.

———. "The Strategy of Ecosystem Development." *Science*, n.s., 164, no. 3877 (1969): 262–70.

Odum, Howard T. *Environment, Power, and Society*. New York: Wiley-Interscience, 1970.

———. *IBP Symposium: Environmental Photosynthesis*. Washington, DC: American Association for the Advancement of Science, 1967.

Oldfield, Frank, Anthony D. Barnosky, John Dearing, Marina Fischer-Kowalski, John McNeill, Will Steffen, and Jan Zalasiewicz. "*The Anthropocene Review*: Its Significance, Implications and the Rationale for a New Transdisciplinary Journal." *Anthropocene Review* 1, no. 1 (2014): 3–7.

Olsakova, Doubravka, ed. *In the Name of the Great Work: Stalin's Plan for the Transformation of Nature and Its Impact in Eastern Europe*. New York: Berghahn Books, 2016.

———. "The International Biological Program in Eastern Europe: Science Diplomacy, Comecon and the Beginning of Ecology in Czechoslovakia (Preprint)." *Environment and History* 24, no. 4 (2018): 543–567.

Oreskes, Naomi, and Erik M. Conway. *Merchants of Doubt: How a Handful of Scientists Obscured the Truth on Issues from Tobacco Smoke to Global Warming*. New York: Bloomsbury Publishing, 2010.

Oreskes, Naomi, and John Krige. *Science and Technology in the Global Cold War*. Cambridge, MA: MIT Press, 2014.

Peeters, Evert, Joris Vandendriessche, and Kaat Wils, eds. *Scientists' Expertise as Performance: Between State and Society, 1860–1960*. London: Routledge, 2015.

Pepper, David, Frank Webster, and George Revill. *Environmentalism: Critical Concepts*. London: Routledge, 2003.

Perkins, John H. *Geopolitics and the Green Revolution: Wheat, Genes, and the Cold War*. Oxford: Oxford University Press, 1997.

Petit, Georges. "Protection De La Nature Et Ecologie." In *International Technical Conference on the Protection of Nature: Lake Success, 22–29 August 1949. Proceedings and Papers*, edited by UNESCO, 304–14. Paris: UNESCO, 1950.

Pickering, Andrew. *Science as Practice and Culture*. Chicago: University of Chicago Press, 1992.

Pielke, Roger A. *The Honest Broker: Making Sense of Science in Policy and Politics*. Cambridge: Cambridge University Press, 2007.

Pistorius, Robin. *Scientists, Plants and Politics: A History of the Plant Genetic Resources Movement*. Rome: International Plant Genetic Resource Institute, 1997.

Plutynski, Anya. "Ecology and the Environment." In *The Oxford Handbook of the Philosophy of Biology*, edited by Michael Ruse, 504–24. Oxford: Oxford University Press, 2008.

Polunin, Nicholas, and Harold K. Eidsvik. "Ecological Principles for the Establishment and Management of National Parks and Equivalent Reserves." *Environmental Conservation* 6, no. 1 (1979): 21–26.

Poole, Robert. *Earthrise: How Man First Saw the Earth*. New Haven: Yale University Press, 2008.

Poore, Duncan. "Meeting on Man and the Biosphere Programme in the Mediterranean Region, Held at Side on the South Coast of Turkey, 5–11 June 1977." *Environmental Conservation* 4, no. 4 (1977): 310–10.

——. "Presentation of Progress Report." In *Proceedings of the Thirteenth (Extraordinary) General Assembly, Geneva 1977*, edited by IUCN, 141–44. Morges: IUCN, 1977.

——. "Progress Report on the Strategy and Its Component Programmes." In *Proceedings of the Thirteenth (Extraordinary) General Assembly, Geneva 1977*, edited by IUCN, 18–20. Morges: IUCN, 1977.

Poore, Duncan, P. Gryn-Ambroes, UNEP, and IUCN. *Nature Conservation in Northern and Western Europe*. Gland: IUCN, 1980.

Poore, Duncan, and IUCN. *Ecological Guidelines for Development in Tropical Rain Forests*. Morges: IUCN, 1976.

Prashad, Vijay. *The Darker Nations: A People's History of the Third World*. New York: New Press, 2008.

Prescott-Allen, Robert. *National Conservation Strategies and Biological Diversity: A Report to IUCN Conservation for Development Centre*. Gland: IUCN, 1986.

Prescott-Allen, Robert, and Christine Prescott-Allen. "Park Your Genes: Protected Areas as in Situ Genebanks for the Maintenance of Wild Genetic Resources." In *National Parks, Conservation, and Development: The Role of Protected Areas in Sustaining Society; Proceedings of the World Congress on National Parks, Bali, Indonesia, 11–22 October 1982*, edited by Jeffrey A. McNeely and Kenton R. Miller, 634–38. Washington, DC: Smithsonian Institution Press, 1984.

Quammen, David. *The Song of the Dodo: Island Biogeography in an Age of Extinctions*. New York: Random House, 2012.

Radkau, Joachim. *Die Ära der Ökologie: Eine Weltgeschichte*. Munich: C. H. Beck, 2011.

——. *Nature and Power: A Global History of the Environment*. Cambridge: Cambridge University Press, 2008.

Raffaelli, David G., and Christopher L. Frid, eds. *Ecosystem Ecology: A New Synthesis*. Cambridge: Cambridge University Press, 2010.

Ray, Carleton. "Promoting 'the Ecology.'" *IUCN Bulletin*, n.s., 13, no. 1/2/3 (1982): 22.

Reid, Walter V., and Kenton Miller. *Keeping Options Alive: The Scientific Basis for Conserving Biodiversity*. Washington, DC: World Resources Institute, 1989.

Reinalda, Bob. *Routledge History of International Organizations: From 1815 to the Present Day*. New York: Routledge, 2009.

Renn, Jürgen, and Manfred D. Laubichler. "Extended Evolution and the History of Knowledge." In *Integrated History and Philosophy of Science: Problems, Perspectives, and Case Studies*, Vienna Circle Institute Yearbook, edited by Friedrich Stadler, 109–25. Vienna: Springer Int'l., 2017.

Renn, Jürgen, Manfred D. Laubichler, and Helge Wendt. "Energy Transformations between Coffee and Coevolution." In *Welcome to the Anthropocene: The Earth in Our Hands; A Special Exhibition by the Deutsches Museum and the Rachel Carson Center for Environment and Society*, edited by Nina Möllers, Christian Schwägerl, and Helmuth Trischler, 79–82. Munich: Deutsches Museum, 2014.

Richardson, David M., ed. *Fifty Years of Invasion Ecology: The Legacy of Charles Elton*. Oxford: John Wiley & Sons, 2011.

Rifkin, Jeremy. *Biosphere Politics: A New Consciousness for a New Century.* New York: Crown, 1991.

Rip, Arie. "Constructing Expertise: In a Third Wave of Science Studies?" *Social Studies of Science* 33, no. 3 (2003): 419–34.

Rist, Gilbert. *The History of Development: From Western Origins to Global Faith.* London: Zed Books, 2014.

Roberts, Sam. "Maurice Strong, Environmental Champion, Dies at 86." *New York Times,* 1 December 2015. Accessed 24 March 2019. http://www.nytimes.com/2015/12/02/world/americas/maurice-strong-environmental-champion-dies-at-86.html?_r=0.

Robertson, Thomas. *The Malthusian Moment: Global Population Growth and the Birth of American Environmentalism.* New Brunswick: Rutgers University Press, 2012.

———. "'This Is the American Earth': American Empire, the Cold War, and American Environmentalism." *Diplomatic History* 32, no. 4 (2008): 561–84.

Roobeek, Annemieke J. "The Crisis in Fordism and the Rise of a New Technological Paradigm." *Futures* 19, no. 2 (1987): 129–54.

Ross, Corey. "Tropical Nature as Global Patrimoine: Imperialism and International Nature Protection in the Early Twentieth Century." *Past & Present* 226, no. 10 (2015): 214–39.

Roston, Michael. "The March for Science: Why Some Are Going, and Some Will Sit Out." *New York Times,* 17 April 2017. Accessed 3 March 2018. https://www.nytimes.com/2017/04/17/science/march-for-science-voices.html.

Rothstein, Robert L. *Global Bargaining: UNCTAD and the Quest for a New International Economic Order.* Princeton: Princeton University Press, 2015.

Rowland, Wade. *The Plot to Save the World: The Life and Times of the Stockholm Conference on the Human Environment.* Madison: Clarke, Irwin, 1973.

Rupert, Mark. *Ideologies of Globalization: Contending Visions of a New World Order.* London: Routledge, 2012.

Sabin, Paul. *The Bet: Paul Ehrlich, Julian Simon, and Our Gamble over Earth's Future.* New Haven: Yale University Press, 2014.

Sachs, Jeffrey D. "From Millennium Development Goals to Sustainable Development Goals." *Lancet* 379, no. 9832 (2012): 2206–11.

Sapp, Jan. *Genesis: The Evolution of Biology.* Oxford: Oxford University Press, 2003.

Sarmiento, Fausto. "Gerardo Budowski: A Beacon to Conservation of Tropical Mountains." *Mountain Research and Development* 22, no. 2 (2002): 197–99.

Sarukhán, José, and Joseph Alcamo. *Ecosystems and Human Well-Being: A Framework for Assessment; A Report of the Conceptual Framework Working Group of the Millennium Ecosystem Assessment.* Washington, DC: Island Press, 2003.

Scheyvens, Regina, Glenn Banks, and Emma Hughes. "The Private Sector and the SDGs: The Need to Move Beyond 'Business as Usual.'" *Sustainable Development* 2016, no. 24 (2016): 371–82.

Schleper, Simone. "Conservation Compromises: The MAB and the Legacy of the International Biological Program, 1964–1974." *Journal of the History of Biology* 50, no. 1 (2017): 133–67.

Schleper, Simone, and Hans Schouwenburg. "Islands and Bioregions: Global Reserve Design Models and the Making of National Parks, 1960–2000." In *Spatializing the His-*

tory of Ecology: Sites, Journeys, Mappings, edited by Raf De Bont and Jens Lachmund, 185–203. New York: Routledge, 2017.

Schmelzer, Matthias. "The Club of Rome to Help the Poor? The OECD, "Development," and the Hegemony of Donor Countries." In *International Organizations and Development, 1945–1990*, edited by Marc Frey, Sönke Kunkel, and Corinna R. Unger, 171–95. London: Palgrave Macmillan, 2014.

Schmidt, Jeremy J., Peter G. Brown, and Christopher J. Orr. "Ethics in the Anthropocene: A Research Agenda." *Anthropocene Review* 3, no. 3 (2016): 188–200.

Schot, Johan, and Vincent Lagendijk. "Technocratic Internationalism in the Interwar Years: Building Europe on Motorways and Electricity Networks." *Journal of Modern European History* 6, no. 2 (2008): 196–217.

Schröder, Iris, and Sabine Höhler. *Welt-Räume: Geschichte, Geographie und Globalisierung sqeit 1900*. Frankfurt am Main: Campus, 2005.

Schulz-Walden, Thorsten. *Anfänge Globaler Umweltpolitik: Umweltsicherheit in Der Internationalen Politik (1969–1975)*. Munich: Oldenbourg Verlag, 2013.

Schuppli, Susan. *Material Witness: Forensic Media Production of Evidence*. Cambridge, MA: MIT Press, forthcoming.

Schwarzenbach, Alexis. *Saving the World's Wildlife: WWF—the First 50 Years*. London: Profile Books, 2011.

Sears, Paul B. "Review: The Careless Technology: Ecology and International Development. The Record of the Conference on the Ecological Aspects of International Development Convened by the Conservation Foundation and the Center for the Biology of Natural Systems, Washington University, December 8–11, 1968, Airlie House, Warrenton, Virginia. Mohammad Taghi Farvar and John P. Milton." *Quarterly Review of Biology* 48, no. 3 (1973): 520–21.

Sellars, Richard West. *Preserving Nature in the National Parks: A History*. New Haven: Yale University Press, 2009.

Sewell, James P. *UNESCO and World Politics: Engaging in International Relations*. Princeton: Princeton University Press, 2015.

Shapin, Steven, and Simon Schaffer. *Leviathan and the Air-Pump: Hobbes, Boyle and the Experimental Life*. Princeton: Princeton University Press, 1989.

Sheail, John. *Nature Conservation in Britain: The Formative Years*. London: Stationery Office, 1998.

———. *Nature's Spectacle: The World's First National Parks and Protected Places*. London: Taylor & Francis, 2014.

———. *Seventy-Five Years in Ecology: The British Ecological Society*. Oxford: Blackwell Scientific, 1987.

Singh, J. P. *United Nations Educational, Scientific, and Cultural Organization (UNESCO): Creating Norms for a Complex World*. London: Routledge, 2010.

Sioli, Harald. "Managing Natural Resources for Scientific, Education and Health Purposes." In *Proceedings of the Twelfth Technical Meeting, Banff, Alberta, Canada 1972* edited by IUCN, 119–228. Morges: IUCN, 1972.

Sismondo, Sergio. "Science and Technology Studies and an Engaged Program." In *The Handbook of Science and Technology Studies*, edited by Edward J. Hackett, 13–32. Cambridge, MA: MIT Press, 2008.

Skyttner, Lars. *General Systems Theory: Problems, Perspectives, Practice*. London: World Scientific Publishing, 2005.

Slack, Nancy G. G. *Evelyn Hutchinson and the Invention of Modern Ecology*. New Haven: Yale University Press, 2010.

Sluga, Glenda. "UNESCO and the (One) World of Julian Huxley." *Journal of World History* 21, no. 3 (2010): 393–418.

Smith, Harold L. *War and Social Change: British Society in the Second World War*. Manchester: Manchester University Press, 1990.

Society of American Foresters. "Obituary: Budowski, Gerardo (1925–2014)." *International Forestry Working Group Newsletter*, December 2014). Accessed 24 March 2019. http://www.orrforest.net/saf/Dec2014.pdf.

Sokolov, Vladimir. "The Biosphere Reserve Concept in the USSR." *Ambio* (1981): 97–101.

Somsen, Geert J. "A History of Universalism: Conceptions of the Internationality of Science from the Enlightenment to the Cold War." *Minerva* 46, no. 3 (2008): 361–79.

Souder, William. *On a Farther Shore: The Life and Legacy of Rachel Carson*. London: Crown Publishers, 2012.

Soulé, Michael E. "What Is Conservation Biology?" *BioScience* 35, no. 11 (1985): 727–34.

Soulé, Michael E., and Bruce A. Wilcox. *Conservation Biology: An Evolutionary-Ecological Perspective*. Sunderland: Sinauer Associates, 1980.

Speich Chassé, Daniel. "Technical Internationalism and Economic Development at the Founding Moment of the UN System." In *International Organizations and Development, 1945–1990*, edited by Marc Frey, Sönke Kunkel, and Corinna R. Unger, 23–45. London: Palgrave Macmillan, 2014.

Spicer, John, and Kevin Gaston. *Physiological Diversity: Ecological Implications*. New York: John Wiley & Sons, 2009.

Stanford University's School of Earth, Energy and Environmental Sciences. "Research," n.d. Accessed 7 November 2016. https://pangea.stanford.edu/ess/research.

Stone, Peter B. *Did We Save the Earth at Stockholm?* Berkeley: Earth Island, 1973.

Stott, Peter H. "The World Heritage Convention and the National Park Service, 1962–1972." *GWS Journal of Parks, Protected Areas & Cultural Sites* 28, no. 3 (2011): 279–90.

Strong, Maurice F. *Canada's Assistance to Developing Nations*. Vienna: Vienna Institute for Development, 1970.

———. "Progress or Catastrophe: Whither Our World?" *Environmental Conservation* 2, no. 2 (1975): 83–88.

———. *Where on Earth Are We Going?* Toronto: Knopf Canada, 2000.

Sullivan, Walter. "Harold Coolidge, Expert on Exotic Mammals." *New York Times*, 16 February 1985.

Suriyakumaran, Canaganayagam. *Environmental Assessment Statements: A Test Model Presentation*. Bangkok: UNEP Regional Office for Asia, and the Pacific and United Nations Asian and Pacific Development Institute, 1980.

Swanson, Timothy. *Global Action for Biodiversity: An International Framework for Implementing the Convention on Biological Diversity*. London: Earthscan, 1997.

Swyngedouw, Erik. "Anthropocenic Promises: The End of Nature, Climate Change and the Process of Post-Politicization." In *Green Utopianism: Perspectives, Politics and Micro-Practices*, edited by Karin Bradley and Johan Hendrén. London: Routledge, 2014.

Takacs, David. *The Idea of Biodiversity: Philosophies of Paradise.* Baltimore: Johns Hopkins University Press, 1996.

Talbot, Lee M. "The World's Conservation Strategy." *Environmental Conservation* 7, no. 4 (1980): 259–68.

Talbot, Lee M., and IUCN. *A Look at Threatened Species: A Report on Some Animals of the Middle East and Southern Asia Which Are Threatened with Extermination.* Cambridge: Fauna Preservation Society for the International Union for Conservation of Nature and Natural Resources, 1960.

Talbot, Ross B. *The Four World Food Agencies in Rome.* Iowa City: Iowa State University Press, 1990.

Tansley, Arthur G. *The British Islands and Their Vegetation.* Cambridge: Cambridge University Press, 1939.

———. "The Use and Abuse of Vegetational Concepts and Terms." *Ecology* 16, no. 3 (1935): 284–307.

———. *The Values of Science to Humanity: The Herbert Spencer Lecture, Oxford University, 2 June 1942.* London: G. Allen & Unwin, 1942.

Terborgh, John. "Preservation of Natural Diversity: The Problem of Extinction Prone Species." *BioScience* 24, no. 12 (1974): 715–22.

Thacher, Peter S. "What's Happened since Stockholm." *Environmental Science & Technology* 8, no. 3 (1974): 214–16.

Thoreau, Henry D. *Walden.* Boston: Houghton, Mifflin and Company, 1882.

Thornton, Thomas F., and Patricia M. Thornton. "The Mutable, the Mythical, and the Managerial: Raven Narratives and the Anthropocene." *Environment and Society* 6, no. 1 (2015): 66–86.

Tilley, Helen. *Africa as a Living Laboratory: Empire, Development, and the Problem of Scientific Knowledge, 1870–1950.* Chicago: University of Chicago Press, 2011.

Tinbergen, Jan, Antony J. Dolman, and Jan van Ettinger. *Reshaping the International Order: A Report to the Club of Rome.* New York: Dutton, 1976.

Tinker, Jon. "World Environment: What's Happening at UNEP?" *New Scientist* 66, no. 953 (1975): 600–3.

Toogood, Mark. "Beyond 'the Toad beneath the Harrow': Geographies of Ecological Science, 1959–1965." *Journal of Historical Geography* 34, no. 1 (2008): 118–37.

Trepl, Ludwig. *Die Idee der Landschaft: Eine Kulturgeschichte von der Aufklärung bis zur Ökologiebewegung.* Bielefeld: Transcript Verlag, 2014.

Trischler, Helmuth. "The Anthropocene: A Challenge for the History of Science, Technology, and the Environment." *NTM Zeitschrift für Geschichte der Wissenschaften, Technik und Medizin* 24, no. 1 (2016): 309–35.

Turchetti, Simone. *Greening the Alliance: The Diplomacy of NATO's Science and Environmental Initiatives.* Chicago: University of Chicago Press, 2018.

Udvardy, Miklos. *A Classification of the Biogeographical Provinces of the World.* Morges: IUCN, 1975.

Uekötter, Frank. *Naturschutz im Aufbruch: Eine Geschichte des Naturschutzes in Nordrhein-Westfalen 1945–1980.* Frankfurt am Main: Campus Verlag, 2004.

———. *Umweltgeschichte im 19. und 20. Jahrhundert.* Munich: Oldenbourg Verlag, 2007.

———, ed. *The Turning Points of Environmental History.* Pittsburgh: University of Pittsburgh Press, 2010.

UN. "Report of the United Nations Conference on the Human Environment. Stockholm, 5–16 June 1972." Accessed 23 March 2019. http://www.un-documents.net/aconf48-14r1.pdf.

———. *Development and Environment (Subject Area V)*. United Nations General Assembly. New York: UN, 1971.

———. *Environmental Assessment of Development Projects*. Kuala Lumpur: UN; Asian and Pacific Development Centre, 1983.

———. "Transforming Our World: The 2030 Agenda for Sustainable Development." Sustainable Development Goals Knowledge Platform, n.d. Accessed 21 March 2019. https://sustainabledevelopment.un.org/post2015/transformingourworld.

UN, and ECOSOC. *Science and Technology for Development: Proposals for the Second United Nations Development Decade; Report of the Advisory Committee on the Application of Science Technology to Development*. New York: UN, 1970.

UNEP. *The Cocoyoc Declaration Adopted by the Participants in the UNEP/UNCTAD Symposium on Patterns of Resource Use, Environment and Development Strategies Held at Cocoyoc, Mexico, from 8 to 12 October 1974*. Nairobi: UNEP, 1974.

———. *Report of the Governing Council of the United Nations Environment Programme on Its 1st Session, Held at the Palais Des Nations, Geneva, from 12 to 22 June 1973*. Geneva: UNEP, 1973.

———. *Taking a Stand: From Stockholm 1972 to Nairobi 1982; Declarations on the World Environment*. New York: UN, 1982.

———. *World Charter for Nature: United Nations Resolution 37/7 of the General Assembly, 28 October 1982*. Nairobi: UNEP, 1982.

UNEP, and UNCTAD. *The Cocoyoc Symposium on "Pattern of Resource Use, Environment and Development Strategies," Cocoyoc, Mexico, October 8–12, 1974*. Zug: Inter Documentation, 1975.

UNESCO. *Biosphere Reserve Nomination Form*. Edited by Division of Ecological and Earth Sciences. Paris: UNESCO, 2013.

———. *Biosphere Reserves: The Seville Strategy and the Statutory Framework of the World Network*. Paris: UNESCO, 1996.

———. "Can We Keep Our Planet Habitable?" *UNESCO Courier* 22, no. 1 (1969): 1–44.

———. *Expert Panel on Project 8: Conservation of Natural Areas and of the Genetic Material They Contain; Final Report*. Paris: UNESCO, 1973.

———. *International Classification and Mapping of Vegetation*. Paris: UNESCO, 1973.

———. "Man and the Biosphere Programme." UNESCO website, n.d. Accessed 24 March 2019. http://www.unesco.org/new/en/natural-sciences/environment/ecological-sciences/man-and-biosphere-programme/.

———. *A Review of the Natural Resources of the African Continent*. Paris: UNESCO, 1963.

———. *Sixty Years of Science at UNESCO, 1945–2005*. Paris: UNESCO, 2006.

———. *Soil Map of the World/6, Africa*. Paris: UNESCO, 1973.

———. *Task Force on Criteria and Guidelines for the Choice and Establishment of Biosphere Reserves, Organized Jointly by UNESCO and UNEP: Final Report*. Paris: UNESCO, 1974.

———, ed. *Intergovernmental Conference of Experts on the Scientific Basis for Rational Use and Conservation of the Resources of the Biosphere, Paris, 4–13 September 1968: Recommendations*. Paris: UNESCO, 1968.

———, ed. *International Technical Conference on the Protection of Nature: Lake Success, 22–29 August 1949: Proceedings and Papers.* Paris: UNESCO, 1950.

UNESCO, MAB, and UNEP, eds. *Ecological Principles for Economic Development: Proceedings of a Seminar and Workshop Held in Blantyre, Malawi, 19–27 May, 1976.* Paris: UNESCO, 1976.

United Nations Conference on the Human Environment. "Declaration on the Human Environment, A/CONF.48/PC.11/Add.42, Stockholm, 16 June 1972" (1972). Accessed 19 March 2019. http://webarchive.loc.gov/all/20150314024203/http%3A//www.unep.org/Documents.Multilingual/Default.asp?documentid%3D97%26articleid%3D1503.

United Nations General Assembly. "2581(XXIV). Recommendation to Convene a United Nations Conference on the Human Environment, A/RES/2581(XXIV), New York, 15 December 1969" (1969). Accessed 19 March 2019. http://www.un.org/ga/search/view_doc.asp?symbol=A/RES/2581%28XXIV%29.

———. "2626 (XXV). International Development Strategy for the Second United Nations Development Decade" (1970). Accessed 24 March 2019. http://www.un.org/en/ga/search/view_doc.asp?symbol=A/RES/2626(XXV).

———. "2997(XXVII). Institutional and Financial Arrangements for International Environmental Co-operation" (1972). Accessed 19 March 2019. https://documents-dds-ny.un.org/doc/RESOLUTION/GEN/NR0/270/27/IMG/NR027027.pdf?OpenElement.

Van der Windt, Henny J. *En dan, wat is natuur nog in dit land? Natuurbescherming in Nederland 1880–1990.* The Hague: Boom, 1995.

Van Dobben, Willem H., and Rosemary H. McConnell, eds. *First International Congress of Ecology.* Wageningen: Centre for Agricultural Publication and Documentation, 1974.

Vanhaute, Eric. "Van Malthus tot Rio: retoriek rond economie en ecologie." *Jaarboek voor Ecologische Geschiedenis* 1 (1999): 77–84.

Vellend, Mark. "Conceptual Synthesis in Community Ecology." *Quarterly Review of Biology* 85, no. 2 (2010): 183–206.

Victor, David G. "Embed the Social Sciences in Climate Policy." *Nature* 520, no. 7545 (2015): 27–29.

Villaume, Poul, Rasmus Mariager, and Helle Porsdam, eds. *The 'Long 1970s': Human Rights, East-West Détente and Transnational Relations.* London: Routledge, 2016.

Vogt, William. *Road to Survival.* Brookfield: W. Sloane Associates, 1948.

Wagar, W. Warren. *A Short History of the Future.* Chicago: University of Chicago Press, 1989.

Walker, Brian. *Ecologists and Environmental Politics: A History of Contemporary Ecology.* Chicago: University of Chicago Press, 1998.

Ward, Barbara. *Spaceship Earth.* New York: Columbia University Press, 1966.

Ward, Barbara, and René J. Dubos. *Only One Earth: The Care and Maintenance of a Small Planet.* New York: W.W. Norton, 1972.

Warde, Paul, Libby Robin, and Sverker Sörlin. *The Environment: A History of the Idea.* Baltimore: John Hopkins University Press, 2018.

———. "Stratigraphy for the Renaissance: Questions of Expertise for 'the Environment' and 'the Anthropocene.'" *Anthropocene Review* 4, no. 3 (2017): 246–58.

Watt, Kenneth E. *Systems Analysis in Ecology.* New York: Academic Press, 1966.

———. "Use of Mathematics in Population Ecology." *Annual Review of Entomology* 7, no. 1 (1962): 243–60.

Weeks, John R. *Population: An Introduction to Concepts and Issues.* Boston: Cengage Learning, 2014.

Weiner, Douglas R. "The Changing Face of Soviet Conservation." In *The Ends of the Earth: Perspectives on Modern Environmental History*, edited by Donald Worster, 252–76. Cambridge: Cambridge University Press, 1988.

Wertz, Wendy R. *Lynton Keith Caldwell: An Environmental Visionary and the National Environmental Policy Act.* Bloomington: Indiana University Press, 2014.

Wessel, Lindzi. "Hundreds Rally for Science at Demonstration near AAAS Meeting." *Science*, 19 February 2017. Accessed 23 March 2019. http://www.sciencemag.org/news/2017/02/hundreds-rally-science-demonstration-near-aaas-meeting.

Westing, Arthur H., and UNEP, eds. *Global Resources and International Conflict: Environmental Factors in Strategic Policy and Action.* Oxford: Oxford University Press, 1986.

Weyler, Rex. *Greenpeace: How a Group of Ecologists, Journalists, and Visionaries Changed the World.* New York: Rodale Books, 2004.

WHO. *National Environmental Health Programmes: Their Planning, Organization, and Administration; Report of a WHO Expert Committee.* Geneva: WHO, 1970.

Wilson, Edward O. *The Future of Life.* New York: Knopf Doubleday Publishing Group, 2002.

Wöbse, Anna-Katharina. "Lina Hähnle (1851–1941): Vogelschutz in drei Systemen." In *Spurensuche: Lina Hähnle und die demokratischen Wurzeln des Naturschutzes*, edited by H.-W. Frohn und Jürgen Rosebrock, 35–56. Essen: Klartext Verlag, 2017.

———. "Oil on Troubled Waters? Environmental Diplomacy in the League of Nations." *Diplomatic History* 32, no. 4 (2008): 519–37.

———. "Phyllis Barclay-Smith: Eine eigensinnige Naturschützerin." *In Vordenker und Vorreiter der Ökobewegung*, edited by Udo E. Simonis, 103–10. Stuttgart: Hirzel, 2014.

———. *Weltnaturschutz. Umweltdiplomatie in Völkerbund und Vereinten Nationen 1920–1950.* Frankfurt am Main: Campus, 2012.

———. "'The World after All Was One': The International Environmental Network of UNESCO and IUPN, 1945–1950." *Contemporary European History* 20, no. 3 (2011): 331–48.

Wong, Laura E. "Relocating East and West: UNESCO's Major Project on the Mutual Appreciation of Eastern and Western Cultural Values." *Journal of World History* 19, no. 3 (2008): 349–74.

Woo, Elaine. "Raymond F. Dasmann, 83: A Founding Father of Environmentalism." *Los Angeles Times*, 9 November 2002. Accessed 18 October 2018. http://articles.latimes.com/2002/nov/09/local/me-dasmann9.

World Commission on Environment and Development. *Our Common Future.* Oxford: Oxford University Press, 1987.

World Resources Institute, IUCN, UNEP, FAO, and UNESCO, eds. *Global Biodiversity Strategy: Guidelines for Action to Save, Study, and Use Earth's Biotic Wealth Sustainably and Equitably.* Washington, DC: World Resource Institute, 1992.

Worster, Donald, ed. *The Ends of the Earth: Perspectives on Modern Environmental History.* Cambridge: Cambridge University Press, 1988.

———. *Nature's Economy: A History of Ecological Ideas; Studies in Environment and History.* Cambridge: Cambridge University Press, 1985.

Worthington, E. Barton. *The Ecological Century: A Personal Appraisal.* Oxford: Clarendon Press, 1983.

————. *The Evolution of IBP*. Cambridge: Cambridge University Press, 1975.

————. "IBP: International Goals." *Science* 161, no. 3839 (1968): 313–14.

Wulf, Andrea. *The Invention of Nature: Alexander von Humboldt's New World*. London: Knopf Doubleday Publishing Group, 2015.

Young, Oran R. "Improving the Performance of the Climate Regime: Lessons from Regime Analysis." In *Oxford Handbook on Climate Change and Society*, edited by John S. Dryzek and Richard B. Norgaard, 625–38. Oxford: Oxford University Press, 2011.

Zalasiewicz, Jan, and Colin Waters. "Media Note: Anthropocene Working Group (AWG)." University of Leicester Press Releases, 29 August 2016. Accessed 23 March 2019. http://www2.le.ac.uk/offices/press/press-releases/2016/august/media-note-anthropocene-working-group-awg.

Index

.

Lightning Source UK Ltd.
Milton Keynes UK
UKHW020931290120
357801UK00006B/279